THE PLANETS

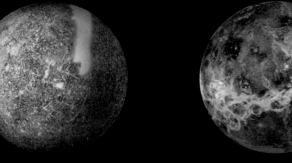

THE PLANETS

A JOURNEY THROUGH THE SOLAR SYSTEM

Giles Sparrow

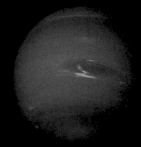

Quercus

Unveiling the Solar System

Our homeworld, the planet we call Earth, is just one of many objects held in gravitational thrall to an average yellow star known as the Sun. The Solar System has members without number: a handful of larger bodies are surrounded by swarms of small worlds and countless chunks of ice and rock – all tracing out their own orbits around the Sun. Traditionally, nine of the larger objects have been known as planets – in order from the Sun they are Mercury, Venus, Earth, Mars, Jupiter, Saturn, Uranus, Neptune and Pluto. However, the precise number of planets has always been open to argument, and the most recent definition has dismissed Pluto from the planetary club, reducing their number to eight.

The nature of the planets

The word planet originates from an ancient Greek word for wanderer: the five planets known to the ancients (excluding Earth and everything beyond Saturn) were merely brighter-than-average 'stars' that moved from week to week against the fixed background of constellations, eventually returning to their original positions, perhaps after many years. All the planets stayed close to the plane marked by the Sun's annual movement through the background stars, so they tended to appear in the same fixed band of star patterns – the constellations of the zodiac.

The obvious interpretation was that the Sun, stars and planets were all moving around Earth at differing speeds, but there were problems with this model from the very start – the rate at which the planets moved through the background stars seemed to vary inexplicably, and sometimes they even stopped in their tracks, retracing their paths in complex loops.

Nevertheless, this Earth-centred model of the Solar System held together, propped up to some degree by its adoption as dogma by the Roman Catholic Church, until the mid-16th century. Around this time, the spirit of independent thinking fostered by the events of the renaissance and reformation combined with some ground-shaking astronomical events to open the way for new theories. Nicolaus Copernicus, an obscure Polish priest, published his famous theory of a Sun-centred Universe in 1543, recognising for the first time that Earth itself was a planet, orbited only by a single satellite, the Moon. However, his theory still had many flaws, and although widely promulgated, there was little in the way of evidence to distinguish it from the ancient Greek model of Ptolemy that had prevailed for so long. Then in 1577, a brilliant comet appeared in Earth's skies, and its path through space was measured for the first time. It was, beyond doubt, a pronounced ellipse, cutting across the orbits of the other planets, and putting an end forever to the idea of planets moving around on crystalline spheres. The passage of the comet inspired the brilliant Danish astronomer Tycho Brahe to a life spent accurately measuring the positions of the planets, and Brahe's measurements enabled his equally brilliant assistant Johannes Kepler to prove once and for all that all of the planets moved in orbits that, while close to circular, were in fact stretched ellipses.

Kepler published his discoveries in the epochal year of 1608, coincidentally the same year as the telescope was invented by Dutch lensmaker Hans Lippershey. Details of this new instrument rapidly spread across Europe, inspiring many scientists and philosophers to build their own versions. The most famous telescopic observer was undoubtedly the Italian physicist Galileo Galilei, who reported a series of staggering discoveries in 1609 and 1610. He found that Venus exhibited phases like the Moon, that the Moon itself had mountains and 'seas', that Jupiter was orbited by four small satellites of its own, and that something very strange was going on around Saturn. With uncanny timing, the conclusive proof that the planets were distant worlds in their own right coincided with the discovery of the means to investigate them in detail.

Alien worlds

Since the early 17th century, the planets have been the most intensively studied of all astronomical objects. As our nearest neighbours in space, they have been discovered anew with each successive generation of improved instruments and revised scientific theories. By the late 1600s, telescopes were powerful enough to allow the structure of the rings around Saturn to be defined, to map dark and bright patches on the surface of Mars, and to trace cloud patterns on Jupiter. The great British scientist Isaac Newton not only invented a new type of telescope, using a mirror instead of a lens for improved performance, but also, in his laws of motion and gravitation, provided the theoretical explanation for the laws of elliptical motion described by Kepler.

New-found precision in the measurements of planetary positions and diameters, coupled with Kepler's laws, allowed the first accurate estimates of the size of the planets, leading to the recognition that there are two types of planet – relatively small 'terrestrial' worlds with diameters of a few thousand kilometres (such as the Earth), and much larger giant planets such as Jupiter and Saturn, many tens of thousands of kilometres across. It was

Cassini captured this Jovian
portrait on its way to Saturn
in December 2000.

Pearl-bright Enceladus
floats above Saturn's rings
and twilight skies.

not until much later, though, that astronomers recognised that the giants were not solid bodies like the inner worlds, but instead were almost entirely composed of gas.

Even as we grew to understand the nature of the planets more, there was still room for surprises. Perhaps the biggest shock of all came in 1781, with William Herschel's serendipitous discovery of a new planet beyond Saturn. It simply hadn't occurred to anyone until that time that there might be distant planets too faint to see with the naked eye, and Uranus doubled the size of the Solar System overnight. Uranus's discovery appeared to confirm a pattern in the distribution of planets (called the Titius-Bode Law) that was only undermined by a distinct gap between Mars and Jupiter. A number of astronomers set out to look for a 'missing' planet in this gap, and in 1801 Ceres, the first of the myriad asteroids was found. These small, rocky worlds, orbiting mostly in a belt between Jupiter and Mars, have a combined volume smaller than Pluto, but their position in the Solar System does indicate a region where Jupiter's enormous gravity affected the pattern of planet formation.

Once the principle that new planets might still await discovery was established, astronomers soon started looking in earnest. Asteroids could only be found by identifying faint 'stars' that moved against the background constellations from night to night, but large planets should be detectable through their gravitational effect on other worlds. In 1846, French mathematician Urbain Le Verrier followed just such a trail of disturbances to predict the location of the eighth planet, Neptune, in the darkness beyond Uranus. Each of the new worlds proved to be a blue-green giant, though both were significantly smaller than Jupiter or Saturn.

All this time, the families of moons around the giant planets continued to grow, surrounding each with its own mini-Solar System of satellites, ranging in size from giants such as Jupiter's Ganymede and Saturn's Titan, both larger than the planet Mercury, down to much smaller worlds at the limit of telescopic observation — most likely captured asteroids such as the two moons of Mars.

Pushing the limits

The discovery of the Solar System's outermost edges, and the multitude of small worlds that inhabit these cold wastes, came in several phases, and it's likely that our picture of the outer limits will change again as technology continues to act as a catalyst to new discoveries.

The first of the outlying worlds to be found was ironically, the result of a deliberate, though misguided, search. Pluto, discovered in 1930 and traditionally catalogued as the ninth planet, remained a lone straggler beyond Neptune for most

of the 20th century, even while theorists suggested it was just the first and brightest among many. Jan Oort proposed that a spherical cloud of icy, dormant comets orbited the Sun at a distance of around a light year, occasionally sending 'long-period' comets falling in towards the warmer climes of the inner Solar System on orbits that last many thousands of years. Gerard Kuiper, meanwhile, suggested that there was a doughnut-shaped ring of 'ice dwarfs' extending from around the orbit of Pluto, within which many of the shorter-period comets, whose orbits last mere decades, reached their furthest point from the Sun.

The indirect evidence for the Oort Cloud is undeniable — which is fortunate since it is well beyond the power of current telescopes or space probes to confirm its existence directly. However, evidence for the Kuiper Belt was far less strong, and it only became an accepted element in most models of the Solar System when the first new Kuiper Belt Objects were discovered in the early 1990s. Within a decade, the numbers had burgeoned, as had the size of objects being discovered — eventually exceeding Pluto itself. This forced astronomers to finally agree a scientific definition of the term 'planet', though the result was controversial and many scientists argue that the ruling that there are eight planets — four terrestrial and four giant) lacks a scientific basis. However, short of defining an equally unscientific arbitrary limit to the size of a planet there seemed to be no way to preserve the status quo.

To the planets

Until the mid-20th century, astronomers could only study the planets as small discs through even the most powerful telescopes. The Space Age that began with the launch of Sputnik 1 in 1957 changed all that. Within a couple of years, the first primitive spaceprobes were lofted beyond near-Earth orbit on the quarter-million-mile journey to the Moon. Each new mission brought a new level of sophistication with it — from the first simple attempts at flybys, through targeted crash-landings, to eventual controlled soft landings and the return of scientific data from the lunar surface.

Even as this armada of probes was paving the way for the manned lunar landings of the US Apollo programme, the first scout missions were venturing much further afield to reconnoitre the inner Solar System, and particularly our immediate neighbours Mars and Venus. The first such probes were limited to brief flybys — there was no way of slowing them down once they reached their destination. Nevertheless, the snapshots of planetary data they returned transformed our understanding of both Mars and Venus (though in the case of Mars this was to be the first of several revisions).

By the 1970s, things had advanced considerably — more powerful rockets and a new technique called the 'gravitational slingshot' allowed spaceprobes to reach the outer planets in reasonable timescales, to rendezvous with inaccessible Mercury, and even visit more than one planet at a time. Meanwhile, improved technology allowed landers to parachute to the surfaces of Venus and Mars. The initial reconnaissance of the Solar System culminated in the hugely successful Voyager probes — a pair of twin spacecraft launched in 1977 on a 'grand tour' of the giant planets. Voyager 1 flew past Jupiter in 1979 and Saturn in 1980, while Voyager 2, following a few months behind, continued from Saturn to Uranus in 1986 and Neptune in 1989.

Aside from the continuing data streaming from the Voyagers, the 1980s were a relatively fallow period in the exploration of the Solar System. Both the Soviet Union and the United States were concentrating on manned spaceflight to Earth orbit, and a number of missions to Mars failed before reaching their destination. The major highlights of the period were the wave of probes sent to rendezvous with Halley's Comet in 1986, and the Magellan mission that produced the first detailed radar maps of Venus from 1990.

Magellan opened the second major wave of exploration, but another mission was already on its way to Jupiter. The Galileo probe was designed to orbit Jupiter for several years, surveying the giant planet and its moons. This was to be followed by the Cassini mission, an even more ambitious Saturn orbiter that reached the ringed planet in 2004. Meanwhile, the jinx on Mars was finally broken in 1997, with the arrival of Mars Pathfinder (a lander that incorporated a small robot rover, a proof-of-concept for the much larger vehicles that trundle over the Martian landscape today), and Mars Global Surveyor, first in an ongoing series of ambitious orbital survey spacecraft.

Elsewhere, a series of missions have targeted the minor worlds of the Solar System — the asteroids and comets. New spacecraft now orbit Venus, and are on their way to opposite ends of the Solar System — fast-moving, baking Mercury and the cold, distant worlds of the Kuiper Belt, including Pluto. There are even ambitious plans for a manned return to the Moon, and eventually a mission to Mars. We are living in a second golden age of space exploration, with new discoveries reshaping our ideas about the Solar System every few months, and a seemingly continuous stream of new missions being designed and launched. It is these missions that have provided most of the stunning images in this book, and to the engineers and planetary scientists that produced them that it is respectfully dedicated.

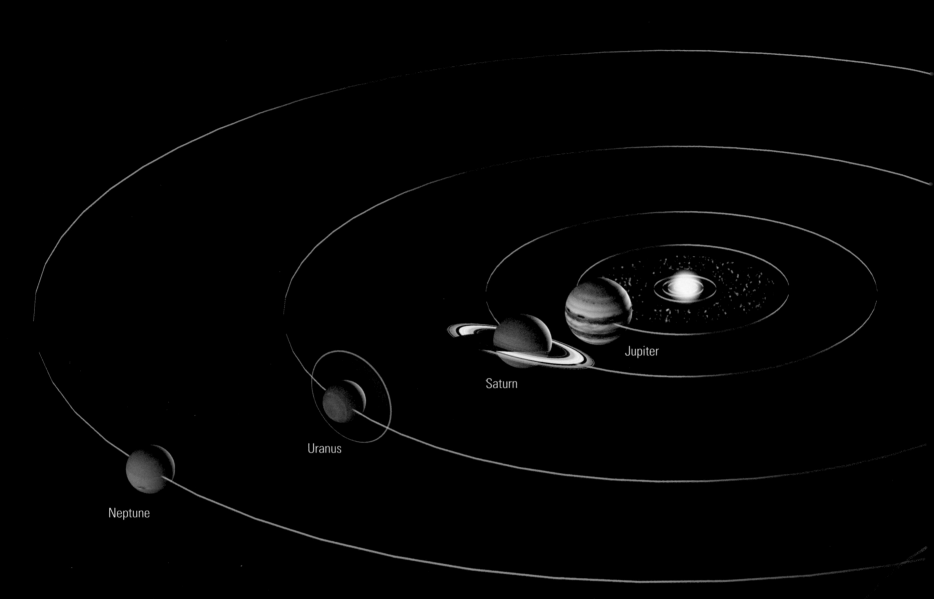

Saturn

Jupiter

Uranus

Neptune

The Solar System

The realm of the planets is divided into two distinct regions by the asteroid belt. The small terrestrial planets stay close to the Sun, orbiting within a few tens of millions of kilometres of each other. The outer giants are separated by hundreds of millions, or even billions of kilometres.

The Sun

Mercury   Venus   Earth   Mars

Jupiter   Saturn   Uranus   Neptune

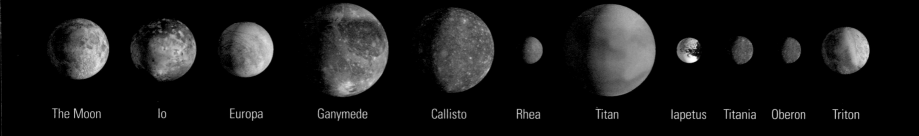

The Moon   Io   Europa   Ganymede   Callisto   Rhea   Titan   Iapetus   Titania   Oberon   Triton

MINOR BODIES

Ceres   Pluto   Charon   2003 UB$_{313}$

Solar System Scales

The Sun dwarfs everything else in the Solar System, and the giant planets are significantly larger than everything else. The diagram opposite shows the Sun and eight planets to scale, while this page shows smaller, but still significant, Solar System bodies to scale with the terrestrial planets.

Probing the Sun

The entire Solar System is dominated by the presence of one object – a huge, roughly spherical ball of gas that contains more than 99 per cent of the entire system's mass. Our local star provides light and heat to all the planets and other objects in this region of space, and has a gravitational grasp that extends for more than a light year in every direction. And yet it is composed mostly of the lightest, simplest elements – hydrogen and helium. Without the Sun, the Solar System would not even exist today.

Astronomers always recognised the Sun as a special type of object, even before they knew what it was. It was often seen as a counterpart to the Moon, due to the remarkable coincidence that makes our planet's satellite and its star appear the same size in Earth's skies. However, even as early as 230 BC, ancient Greek astronomers such as Aristarchus of Samos had estimated the distance to the Sun, revealing that it was a truly massive, distant object compared to the relatively small but nearby Moon. If this knowledge had become more widespread, then the cosmology of Ptolemy, in which the Sun, moon, stars and planets all circled the Earth, might have lost its grip long before the time of Copernicus, Kepler and Galileo.

The Sun's brilliance in our skies has always been a positive disadvantage to those seeking to study it. The bright disc that we usually see, called the photosphere, simply marks the region where the Sun's gases grow sparse enough to become transparent. Here, energy that may have spent tens of thousands of years forcing its way out from the core, is finally released as visible light and a host of other radiations.

From Earth, the Sun's outer layers are only ever briefly visible during rare total eclipses, when the Moon passes directly in front of the Sun and blocks out the photosphere. A wealth of detail now appears around the black disc of the Moon – reddish loops called prominences, the occasional much brighter solar flare, and the milky-white light of the corona, the Sun's outer atmosphere. Temperatures here are far higher than in the photosphere – up to 1 million °C (1.8 million °F),

yet the gas is so sparse that its light is invisible alongside the photosphere. On Earth, there's another problem – our atmosphere is so dense that molecules 'scatter' the intense light from the photosphere, bouncing light rays onto new paths. The phenomenon is strongest for shorter-wavelength blue light, which is why sunlight gives the entire sky a bluish glow.

Since the atmosphere has such a marked effect on radiation from the Sun, it's small wonder that the first instruments sent above the atmosphere completely changed our view of it. For the first time, detectors found highly energetic gamma rays, X-rays and extreme ultraviolet radiation emerging from the Sun. It also became clear that Earth was moving through a constant blizzard of high-speed, high-energy particles blowing out from the corona. The occasional solar flares ejected even more material into this 'solar wind'. Where the particles met the magnetic field around Earth or any other planet, they rained down around the magnetic poles as colourful, glowing aurorae.

For most observations of the Sun, Earth orbit has proved an ideal vantage point. Situated above the atmosphere, a solar telescope can detect all the hidden, non-visible radiations normally soaked up in the upper air. Here, too, a simple opaque 'occulting disc' can hide the light of the photosphere and reveal the Sun's fainter surroundings. Satellites such as SOHO and TRACE, whose images feature in the following pages, are able to do their work from Earth orbit, but a few spacecraft have travelled further afield to study the Sun from new angles. Pioneers 5–9 were 1960s NASA probes put into orbit around the Sun to record the solar wind, while Ulysses was an ambitious 1990s mission that entered a polar orbit which took it far above and below the Sun, allowing it to study the normally hidden polar regions.

The Sun's core occupies a relatively small region at the centre. It is surrounded by the radiative zone, a transparent but 'foggy' region of hot gas, then the opaque convective zone, capped by the photosphere.

Distance from Sun	Orbital period	Orbital eccentricity	Diameter	Surface gravity	Rotation period	Axial tilt	Natural satellites
0	–	–	1.4	270	29	0	–
million km	Earth days		million km	g	days (average)	degrees	

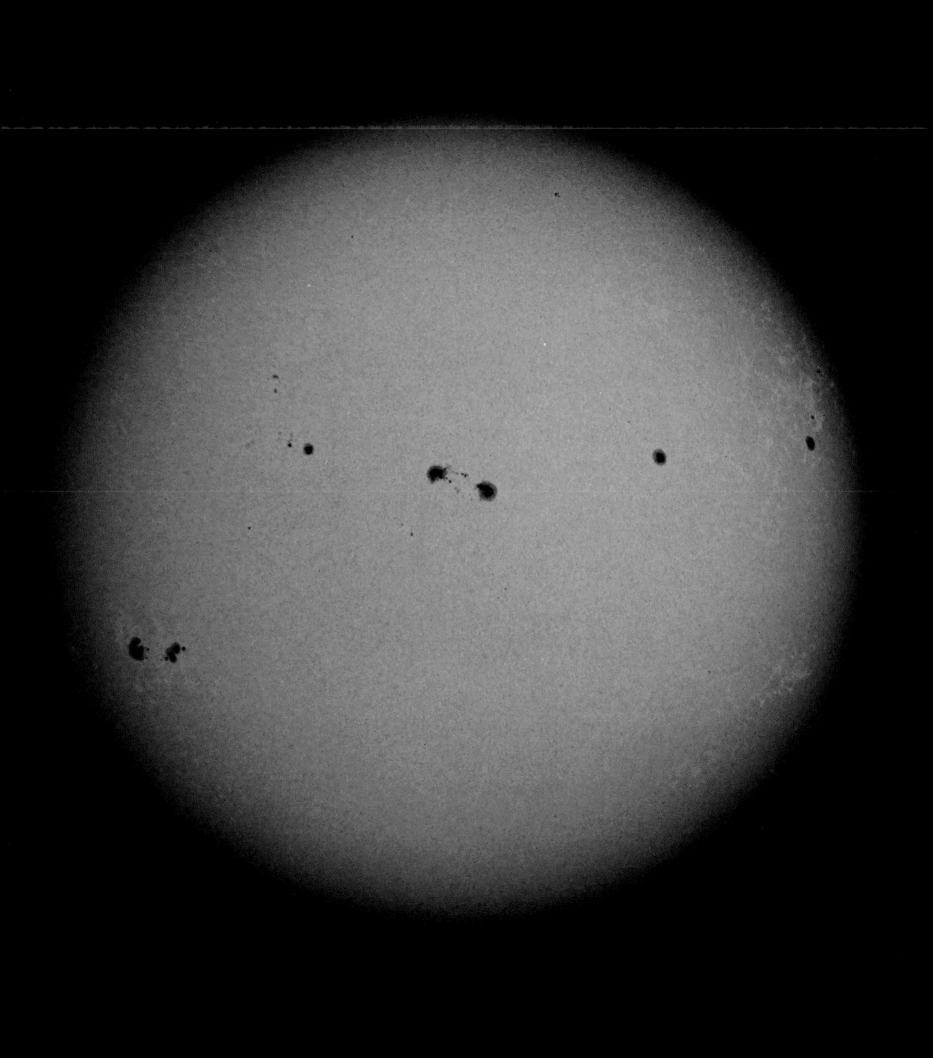

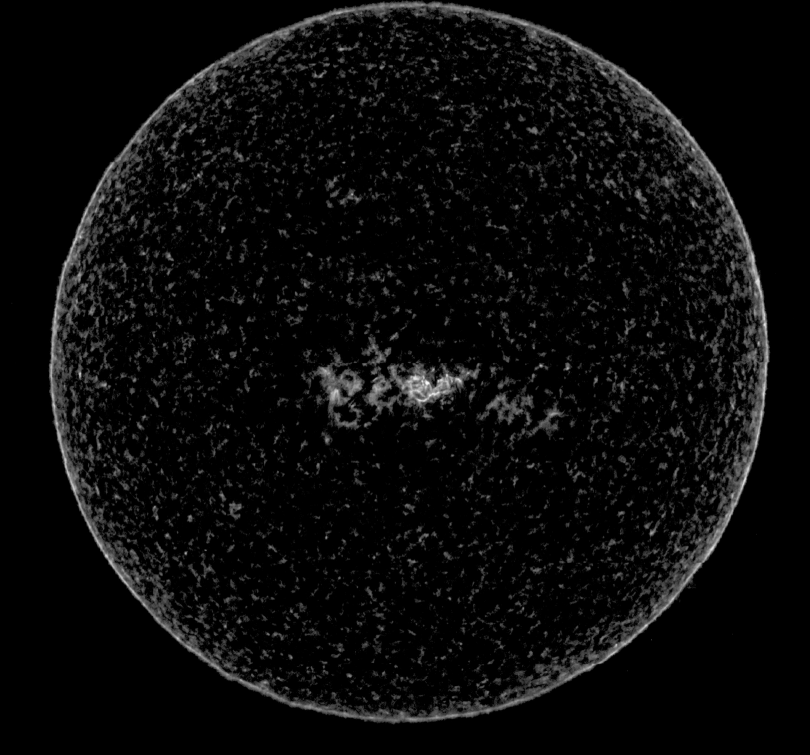

149.6

million km
from the Earth

The Sun

Solar granulation

Surface feature

An ultraviolet image tuned to radiation at 20,000 °C (36,000 °F) reveals a seething mass of dark cells, a pattern called granulation. Each granule, about 10,000 km (6,000 miles) across, is the top of a convection cell that stretches deep into the solar interior. Hot gas rises from closer to the core, releases its energy at the photosphere, cools and falls back around the edges of the cell.

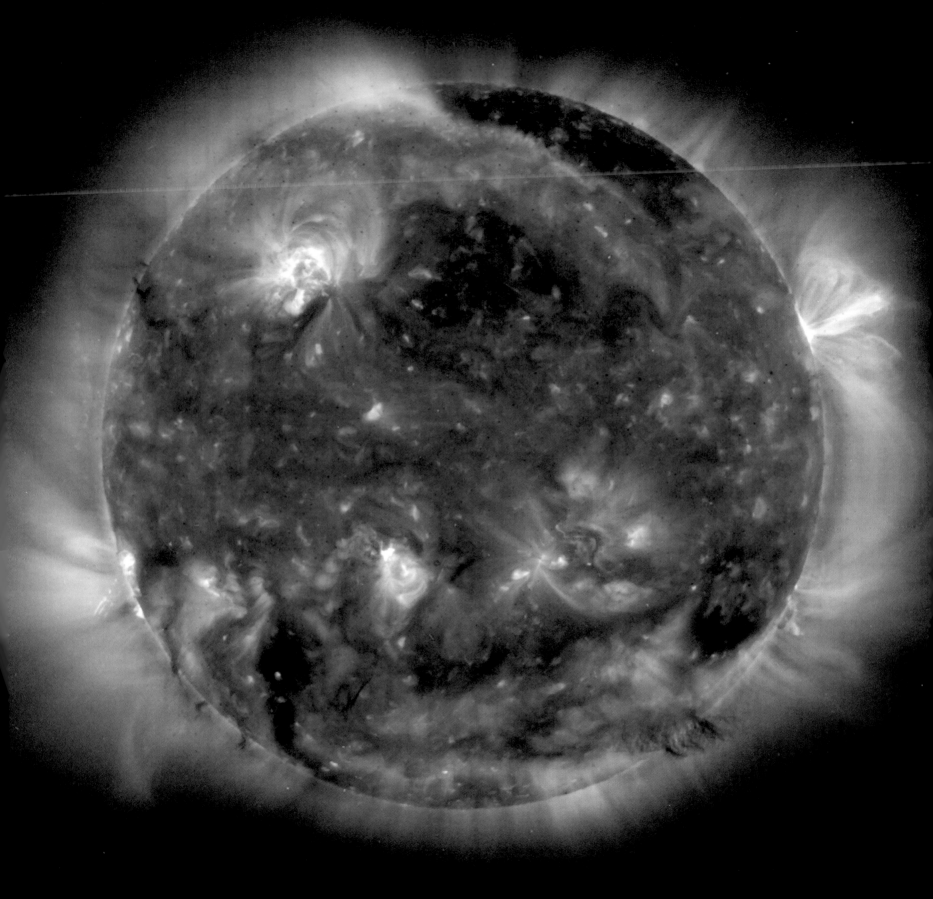

149.6

million km
from the Earth

The Sun

Extreme Sun

Magnetic features

This image from the Solar Heliospheric Observatory (SOHO), captures a number of different radiations, revealing gas at different temperatures. The spectacular result shows how complex the apparently placid Sun can become, its surface and outer atmosphere a tangle of twisted gas streamers, guided by the invisible but powerful magnetic field emerging from within.

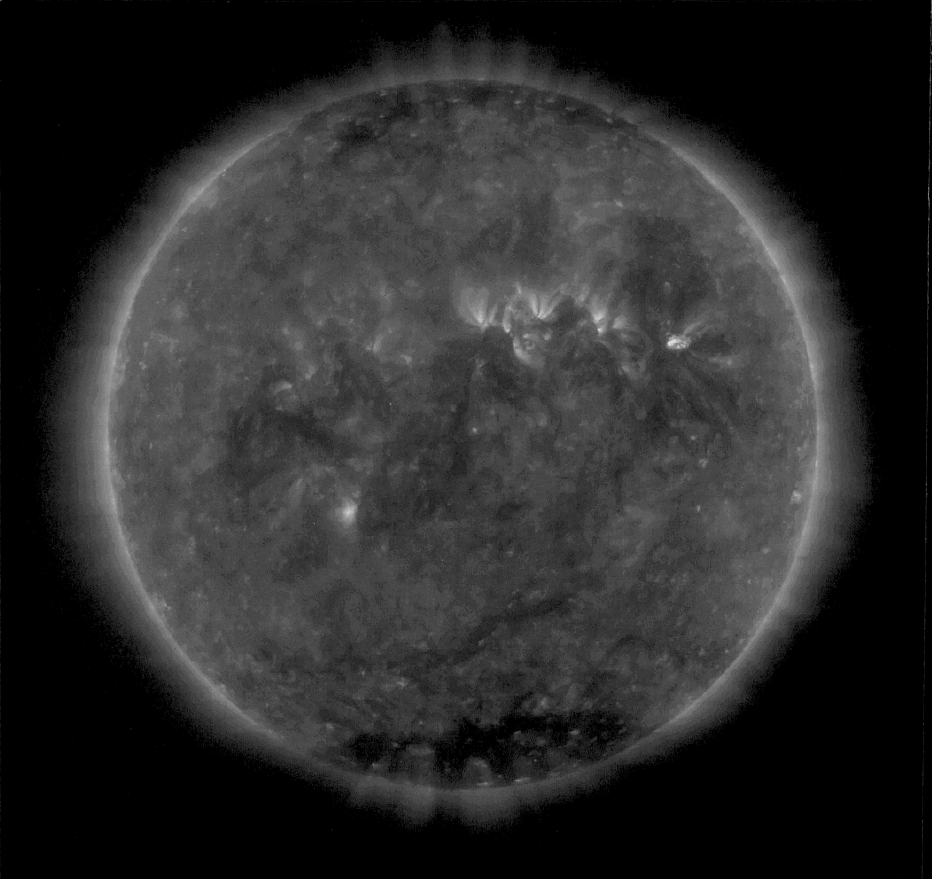

149.6

million km
from the Earth

The Sun

The quiet Sun

Ultraviolet features

An extreme ultraviolet image from SOHO shows high-temperature regions of the Sun. This image was taken during the quiet phase of the Sun's 11-year solar cycle. At this time, the magnetic field is at its weakest, and so there is little to disturb the Sun's surface or disrupt the passage of heat and light from the core into space through the granular convection cells.

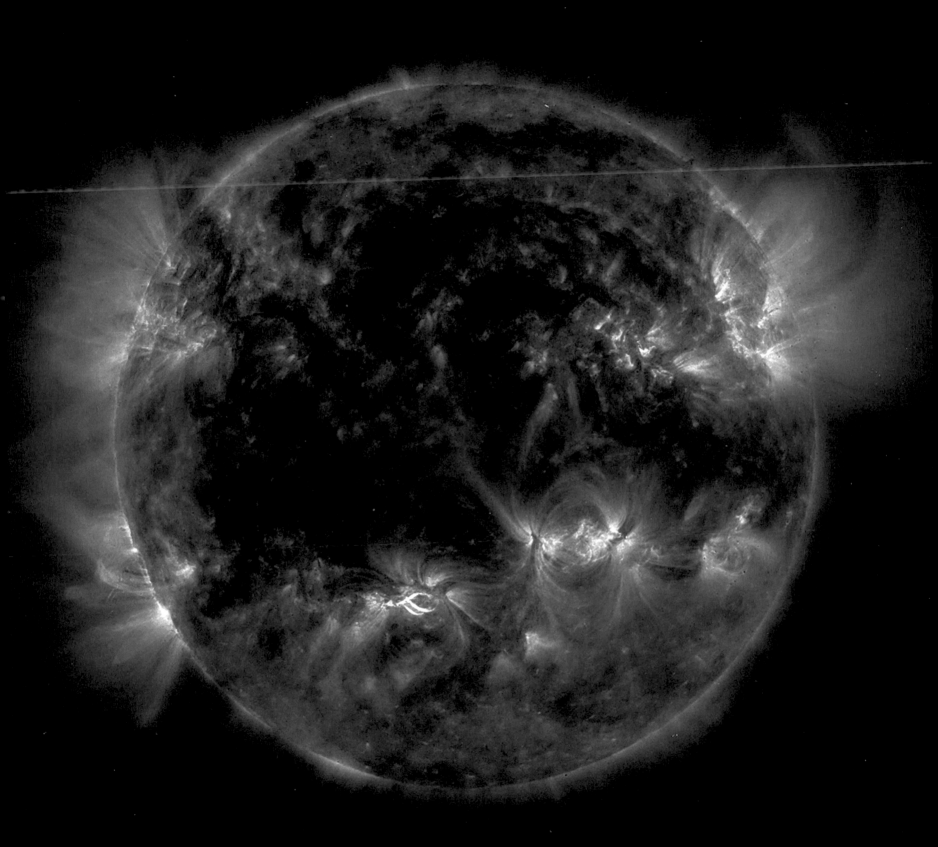

149.6

million km
from the Earth

The Sun

The active Sun

Ultraviolet features

Two years later, a similar image tells a very different story. Because the Sun is not solid, different regions rotate at different rates, causing the magnetic field embedded within to become tangled, and burst out from the photosphere. The Sun's overall energy output rises only slightly, but the magnetic field acts like a dam, occasionally letting huge bursts of energy escape in solar flares.

149.6

million km
from the Earth

The Sun

Solar prominence

Atmospheric feature

Prominences are bright arcs of material looping above the solar surface. They show where hot gas flows along loops of the solar magnetic field through the 'transition region' between the cool but visible photosphere and the hot but sparse corona. Where a magnetic loop 'short circuits' at a lower level in the atmosphere, it can release huge amounts of energy, generating solar flares.

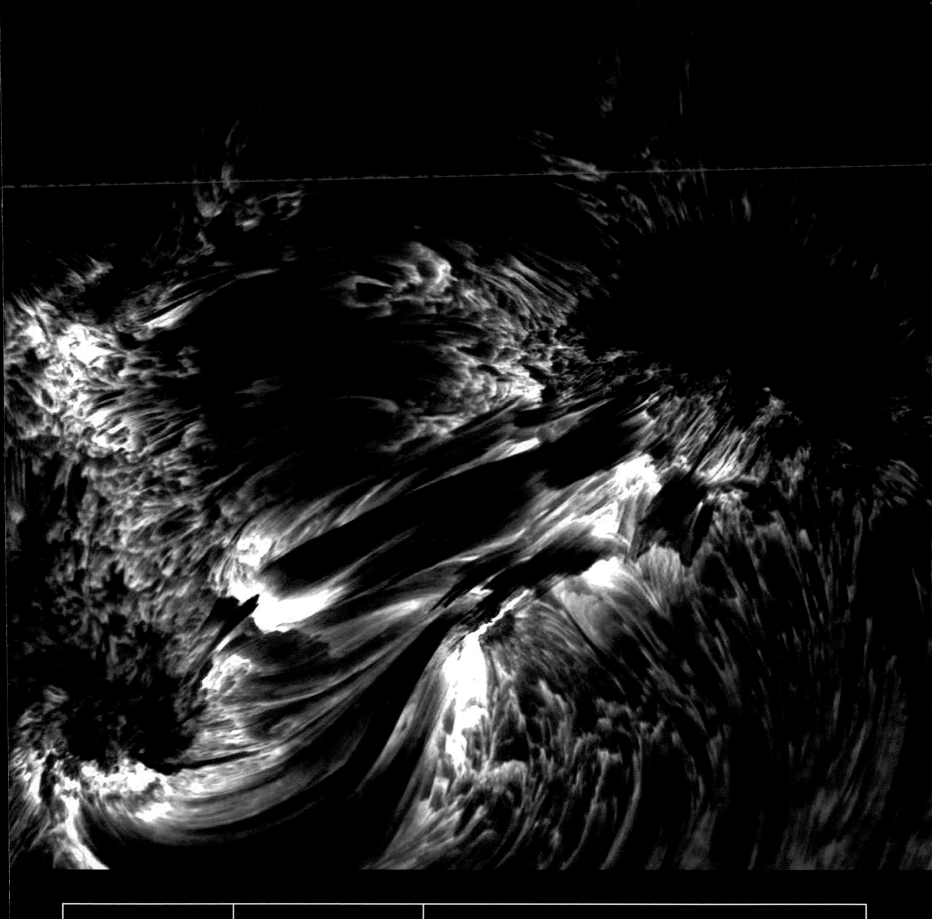

149.6

million km
from the Earth

The Sun

Spicules

Atmospheric feature

The apparently smooth surface of the Sun is in fact alive with activity even during quiet phases. 10,000-km (6,000-mile) spicules snake out from the surface – towering, threadlike pillars of flame that transfer heat from the visible surface of the photosphere to the invisible, sparse, but extremely hot corona. Here, thread-like spicules mark the fringes of a dark, cool sunspot.

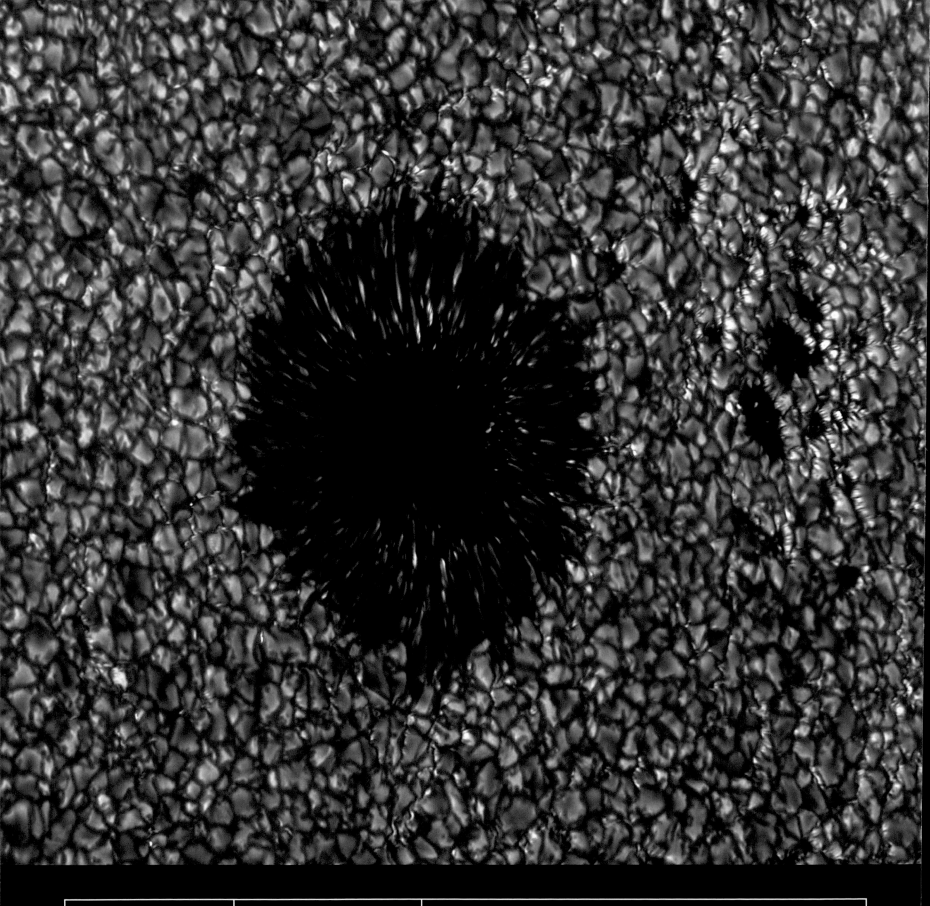

149.6

million km
from the Earth

The Sun

Regular sunspots

Surface feature

Sunspots are the most prominent of the Sun's visible surface features, large dark regions where magnetic disturbances create cooler clearings in the solar gases. Despite temperatures of around 3,500 °C (6,300 °F), they appear dark and cool compared to surroundings about 2,000 °C (3,600 °F) hotter. Individual spots are frequently larger than the Earth.

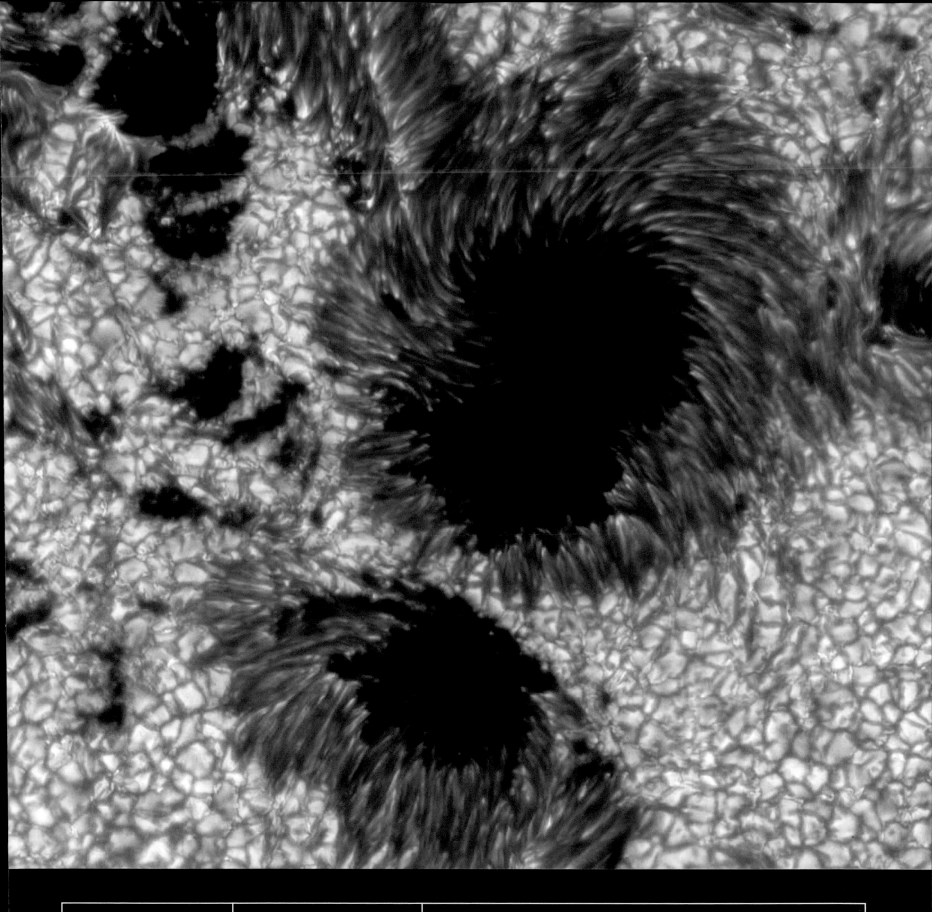

million km
from the Earth

The Sun

Irregular sunspots

Surface feature

Sunspots normally come in pairs with opposite magnetic properties – one marks the point where a loop of magnetic field breaks out through the photosphere, another marks the point where it re-enters. However, around the peak of each solar cycle, the Sun's magnetic field becomes so twisted that it gives rise to far more complex patterns and groups.

149.6

million km
from the Earth

The Sun

Coronal Mass Ejection

Atmospheric feature

The most spectacular solar events are Coronal Mass Ejections (CMEs), vast waves of material that burst from the Sun's atmosphere and roll across the Solar System, carrying an intense magnetic field within them. As CMEs pass Earth, they disrupt our own planet's magnetic field, triggering bright aurorae, damaging satellite electronics and sometimes even causing power blackouts.

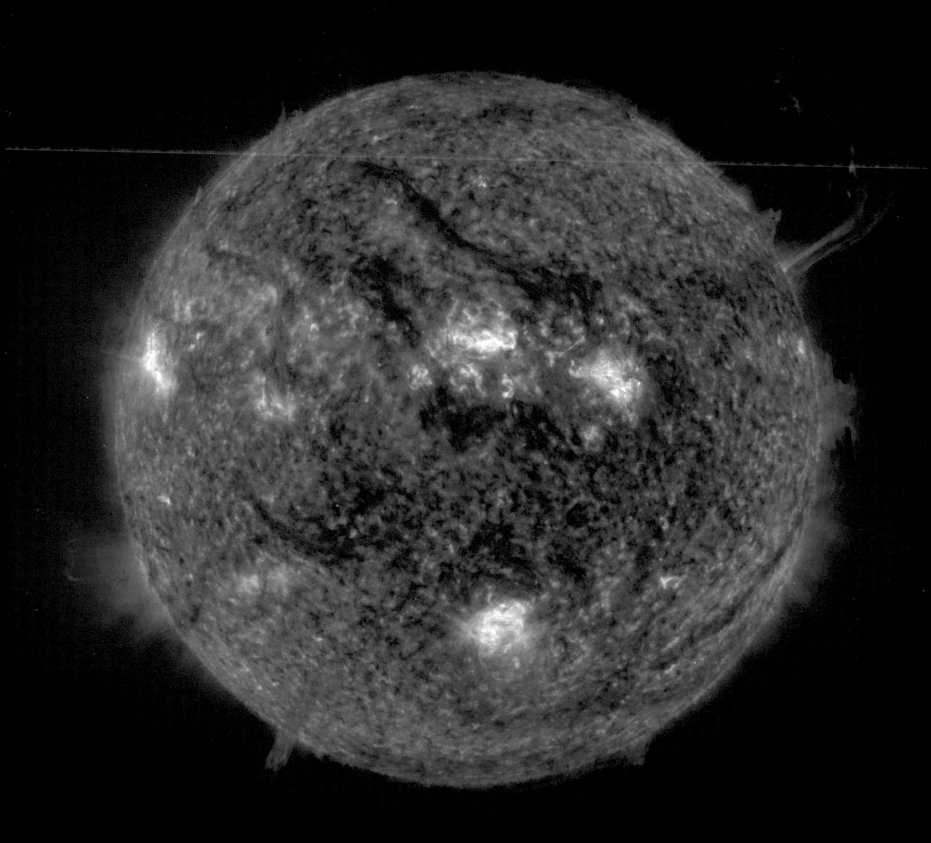

149.6
million km
from the Earth

The Sun

Solar flare

Atmospheric feature

This image from the Transition Region and Coronal Explorer (TRACE) satellite shows a huge solar flare above the Sun's surface. Flares originate lower in the corona than CMEs, but often share the same trigger. When magnetic field loops pushing far out of the Sun suddenly 'short circuit' at a lower level, enormous amounts of energy are released and poured into the solar atmosphere.

Racing Mercury

The closest planet to the Sun is also one of the most obscure, and very little was known about it until well into the Space Age. Even today, it's still the terrestrial world of which we know the least – though hopefully at least some of the gaps in our knowledge will be filled in the next few years.

Mercury is a double challenge to Earthbound observers because of its comparatively tiny size and its proximity to the Sun. With a year lasting just 88 Earth days, Mercury speeds around the Sun and never gets far away from it in Earth's skies. As a result, it only ever appears in the twilight skies before dawn or after sunset, and can only ever be seen through a thick layer of atmosphere – never in a clear, dark sky.

Although some optimistic astronomers drew up maps based on their observations of Mercury before the Space Age, the features they recorded bore little relationship to the planet eventually photographed during spaceprobe flybys. In fact, the only real features of Mercury that can be seen from Earth are its changing phases – the Moon-like sequence which ensures that when Mercury is fully illuminated by the Sun, it is also at its most distant from us on the far side of its orbit and completely impossible to observe.

Sending a spaceprobe to Mercury is also a unique challenge, though. The laws of planetary motion mean that planets closer to the Sun move along their orbits much faster than those further away – so while Earth moves at roughly 30 km/s (19 miles per second), Mercury moves at around 48 km/s (30 miles per second). To catch up with Mercury and have any hope of getting into orbit around it, a spacecraft needs to pick up a great deal of extra speed, and this a challenge for even the latest generation of rockets.

In order to send the first probe to Mercury, NASA's engineers had to take a shortcut. They could not hope to give Mariner 10 enough speed to match Mercury's orbit, but they could boost it enough to reach an elliptical orbit lasting 176 days, or two Mercury years, and intersecting Mercury's own orbit at its closest to the Sun. By dropping their probe into an orbit of this kind, going the opposite

way around the Sun, they could arrange it so that Mariner 10 would rendezvous with Mercury at 176-day intervals. This was how the probe eventually made three flybys of Mercury between March 1974 and March 1975, returning photographs that revealed a heavily cratered, Moon-like world that nevertheless shows signs of its own distinct and unusual history.

There was one major catch with the Mariner 10 approach, though. Tidal forces from the Sun have acted to slow Mercury's rotation in a similar way to how Earth has slowed the Moon. In this case, though, Mercury does not have a day that matches its year – rather it rotates in precisely two thirds of a Mercury year, or 58.7 Earth days. As a result, during each of its flybys Mariner 10 found Mercury in exactly the same orientation, with the same side of the planet illuminated, and the other half hidden in frustrating darkness.

Despite representing an obvious gap in our knowledge of the inner Solar System, a return to Mercury took a long time to become a priority for planetary scientists. One of the reasons this finally changed was the discovery of possible ice at Mercury's poles. Despite the fact that the majority of the planet's surface is baked to temperatures of around 420 °C (790 °F) by the searing sunlight, there are a few craters at the poles that lurk in permanent shadow, and where ice dumped by long-ago comet impacts might still survive.

Now a new probe is finally on its way – the MESSENGER (Mercury Surface, Space Environment, Geochemistry and Ranging) mission was launched in 2004, and will finally enter orbit around Mercury in 2011. In order to match speed with Mercury for this orbital feat, it is taking a remarkably long journey, involving two manoeuvres round Earth, two more around Venus, and three flybys of Mercury itself as it slowly spirals sunward. When it finally reaches orbit and begins to return scientific data, our understanding of Mercury will be undoubtedly transformed beyond recognition.

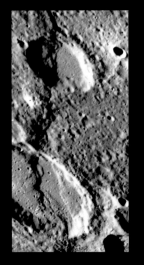

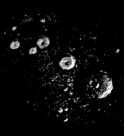

Discovery Rupes is just one of many huge cliff systems that wind their way across Mercury's surface, separating neighbouring regions with sheer drops and rises of up to a kilometre. The rupes are thought to show where Mercury once shrank, and parts of its crust popped outwards in order to fit.

This image shows craters at Mercury's pole, coinciding with a region that produces bright radar reflections. Because Mercury orbits almost bolt upright, these craters receive no direct sunlight and so could possibly shelter deposits of water ice.

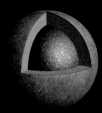

Mercury's strange internal structure seems to be dominated by an enormous core of solid iron that increases the planet's density. This is surrounded by a mantle and thin crust of silicate rocks.

Distance from sun	Orbital period	Orbital eccentricity	Diameter	Surface gravity	Rotation period	Axial tilt	Natural satellites
57.9	88	0.205	4,875	0.38	59	0.01	0
million km	Earth days		km	g	Earth days	degrees	

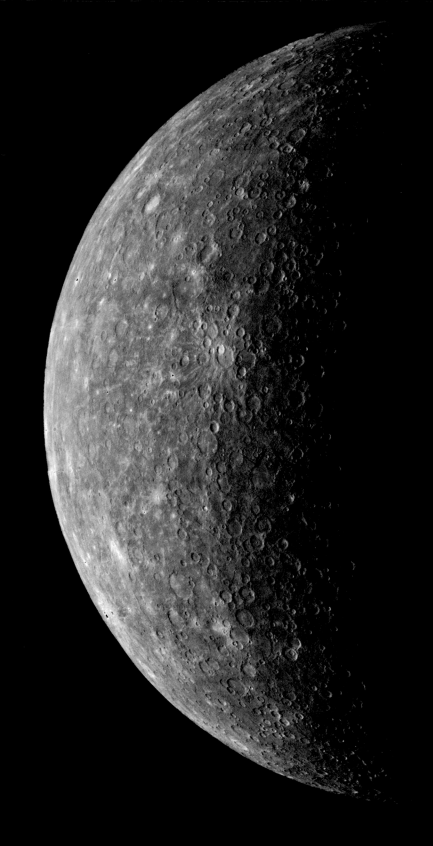

57.9

million km
from the Sun

Mercury

Mercury's limb

Terrestrial planet

Mercury is the smallest major planet – at 4,875 km (3,028 miles) across, it is only slightly bigger than Earth's Moon, and smaller than some large satellites of the outer planets. Yet Mercury is the densest planet for its size – its metal core is not much smaller than Earth's, suggesting that much of its outer mantle and crust were stripped away, probably by a collision, early in its history.

57.9

million km
from the Sun

Mercury

Caloris Basin

Impact basin

Four billion years ago, Mercury suffered one of the largest impacts in the history of the Solar System, as a large asteroid slammed into it at tremendous speed. The shockwaves devastated an entire hemisphere, and hot molten rock from the interior burst up through the pulverised surface to heal the wound. The result is an enormous impact structure known as the Caloris Basin.

Lifting the veil on Venus

Earth's nearest planetary neighbour is also the brightest object in our skies apart from the Sun and Moon. At its closest approach, its diameter is more than 1/30 the size of the full Moon. The planet's changing phases, caused as different portions of the illuminated side become visible from Earth, are clear through even the smallest telescope, and yet Venus remained an obstinate, impenetrable mystery until the beginning of the Space Age. For Venus is cloaked in a thick atmosphere that completely obscures its surface while bouncing back 65 per cent of the sunlight that strikes it.

With close approaches that bring it to within 41 million km (26 million miles) of Earth, Venus was an obvious target for early planetary exploration. The Soviet Union launched the first in its series of Venera probes towards Venus in 1961, but a series of malfunctions meant that NASA's Mariner 2 became the first spacecraft to complete a Venus flyby in December 1962. The probe sent back the first close-up pictures of Venus, revealing only that its atmosphere was just as opaque at close range as it was at a distance. However, analysis of light from the atmosphere did show that its composition was dominated by carbon dioxide, and that the temperatures were extremely high.

From the mid-1960s onward, the Soviet probes began to have more luck. Their aim now was to land on Venus and send back pictures from the surface, and to use radar to pierce through the clouds and map the terrain below. But although it was now clear that the Venusian atmosphere and surface were hostile, the Soviet engineers still underestimated just how hostile they were. Veneras 4, 5 and 6 failed during their descent through the upper atmosphere, and Venera 7 was probably the first probe to reach the surface. By the time it landed, however, the signals were so scrambled that no picture could be returned – it was only later analysis that revealed the probe had managed to send back readings of constant temperature and pressure, indicating that it had stopped descending through the atmosphere. The Venusian surface, it appeared, had a crushing atmosphere with 100 times the pressure of Earth's, and a temperature of around 475 °C (890 °F). Later, more heavily armoured Veneras were able to confirm these conditions, and finally send back pictures of the planet's surface – a landscape of baked volcanic rock.

Clearly radar held the key to understanding Venus in detail, and the early Venera orbiters were followed in the 1970s and 1980s by NASA's Pioneer Venus mission, and eventually by the far more sophisticated Magellan (1989-1994). While early radar maps had been able to distinguish a few areas of high and low terrain (most notably the Maxwell Montes highland region near the equator), Magellan applied techniques developed for Earth remote sensing satellites, and was able to map smaller regions of the planet with much greater detail. Magellan's data could reveal not only the height of terrain below it, but also its surface slope and roughness or smoothness. It could even distinguish between regions with different mineral make-up. Many of the images on the following pages result from combining these different types of radar data to produce three-dimensional images, allowing us to 'see' individual features and whole areas of the Venusian landscape that will probably remain forever hidden from the direct view of our atmospheric probes and landers.

After a long gap in the exploration of Venus, a new European probe entered orbit around the planet in 2006. Venus Express will study the planet using sophisticated cameras rather than radar. It may not produce three-dimensional images like those of Magellan, but it will be able to measure temperature variations on the surface (perhaps indicating volcanic activity), and is already revealing previously hidden features of the Venusian weather.

This historic first colour image, sent back to Earth from Venera 13 in 1982, reveals a brownish landscape scattered with plates of broken volcanic rock. The probe functioned for just 127 minutes on the hostile Venusian surface.

Venus is almost the same size as the Earth, and has a quite similar internal structure, with a thin silicate crust, a deep, rocky mantle and a core of nickel and iron. However, more of Venus's core has frozen solid.

Distance from sun	Orbital period	Orbital eccentricity	Diameter	Surface gravity	Rotation period	Axial tilt	Natural satellites
108.2	224.7	0.006	12,104	0.9	243	177.36	0
million km	Earth days		km	g	Earth days	degrees	

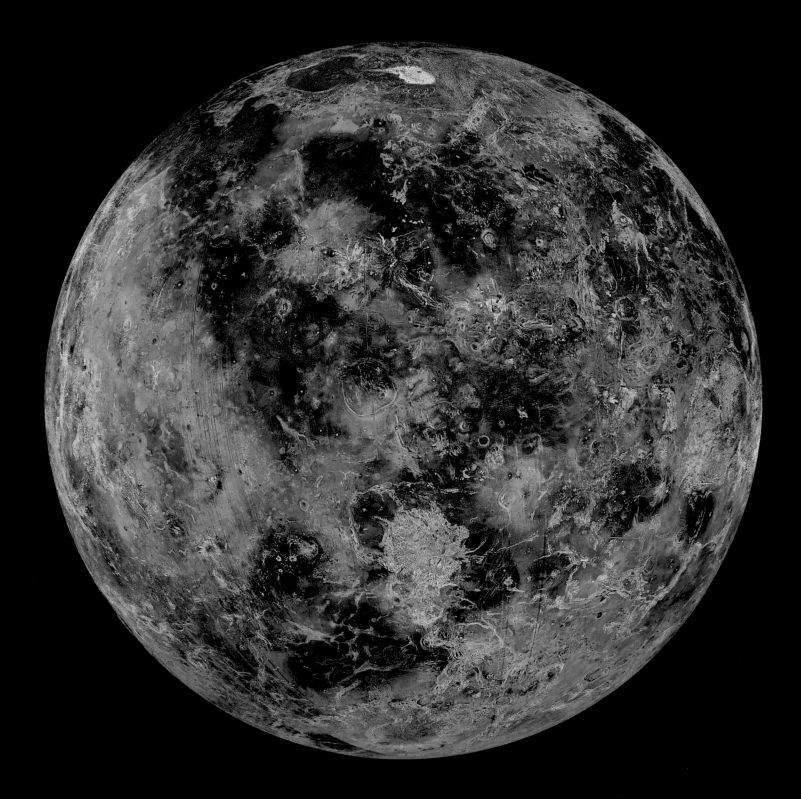

108.2

million km
from the Sun

Venus

Venus globe 0°

Terrestrial planet

Radar maps are the only way to view surface details through the opaque Venusian atmosphere. The sequence of four globes on this and the following pages is based on data from the Magellan, Pioneer Venus and Venera orbiters. Towards the top of this view, centred on longitude 0°, are the highlands of Ishtar Terra, including the Maxwell Montes mountains, 11 km (7 miles) high.

108.2

million km
from the Sun

Venus

Venus globe 90°

Terrestrial planet

A view of the Venusian globe centred on longitude 90° is dominated by the large highland region known as Aphrodite Terra. This region, roughly the size of Africa, is home to many of the best-known Venusian mountains. On these false-coloured maps, blue represents comparatively low-lying regions, green intermediate terrain, and brown and white the Venusian highlands.

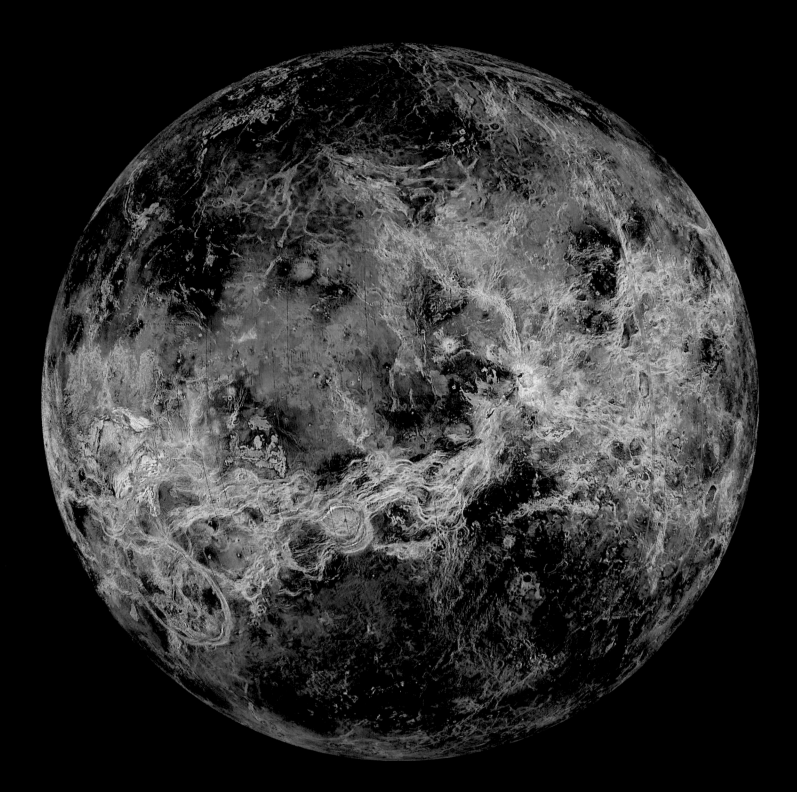

108.2

million km
from the Sun

Venus

Venus globe 180°

Terrestrial planet

A view centred on 180° longitude clearly shows the deep canyons of 'chasmata' that wind and loop their way around much of the Venusian equator. Although Venus today does not appear to have active tectonic plates like those on Earth, these scars may be a sign that once, in its ancient history, such plates began to develop, but soon ground to a halt.

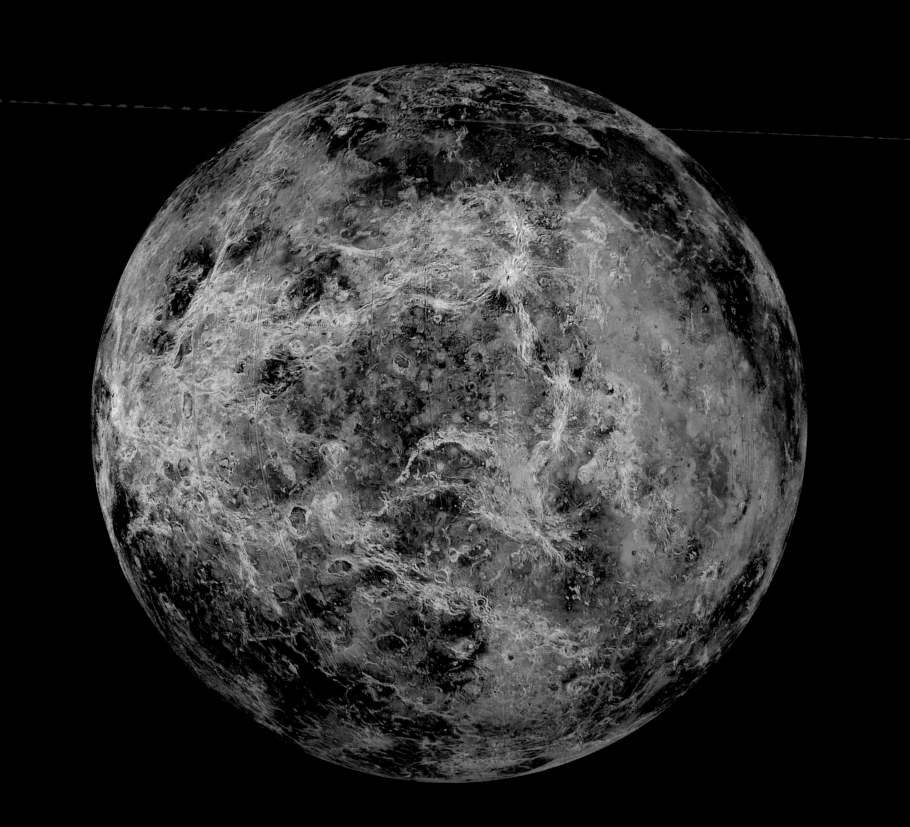

108.2

million km
from the Sun

Venus

Venus globe 270°

Terrestrial planet

Most of the Venusian surface consists of low-to-average height terrain – the highlands are very distinct. By counting the number of impact craters formed in the lowland areas compared to those on the higher plateaus, astronomers have shown that the lowland regions (and even some of the highlands) were wiped clean by widespread volcanic eruptions about 500 million years ago.

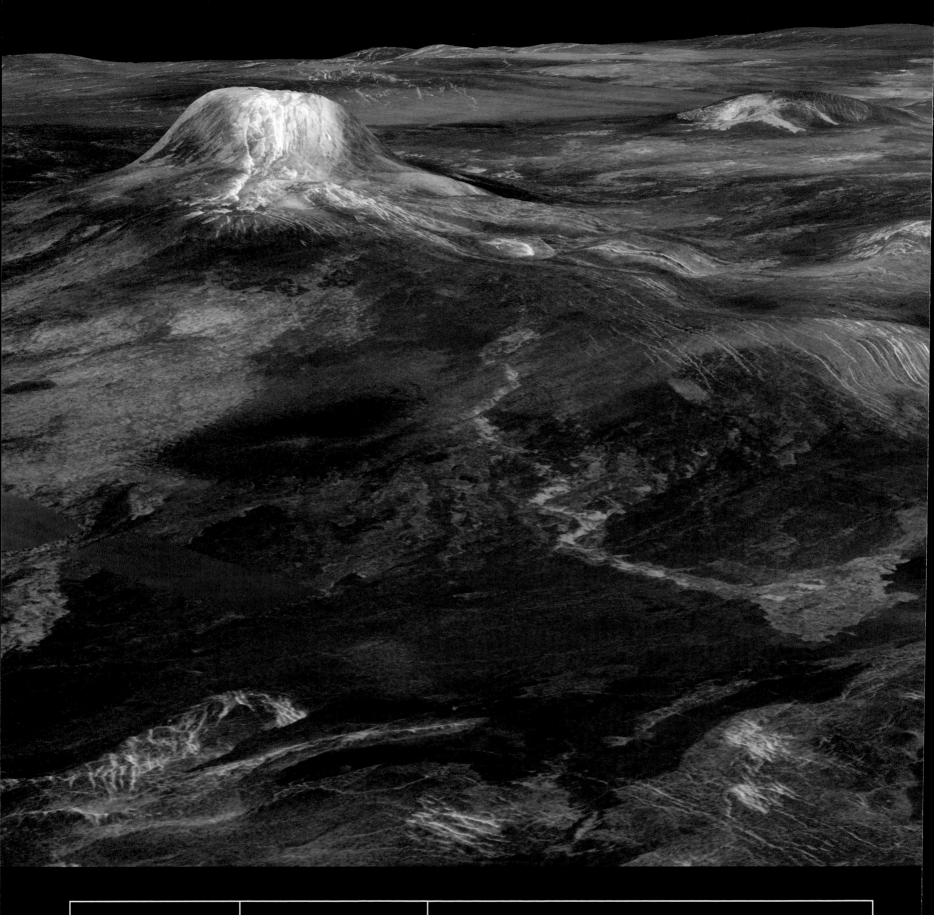

108.2

million km
from the Sun

Venus

Eistla Regio

Volcanic plain

Reconstructions of the Venusian surface based on Magellan data reinforce the fact that Venus is a volcano world – nearly every feature can be traced back to a volcanic origin. This view shows plains of solidified lava around Gula Mons (left) and Sif Mons (right). The vertical scale is exaggerated to reveal variations in height, but in reality these volcanic mountains still rise for many kilometres.

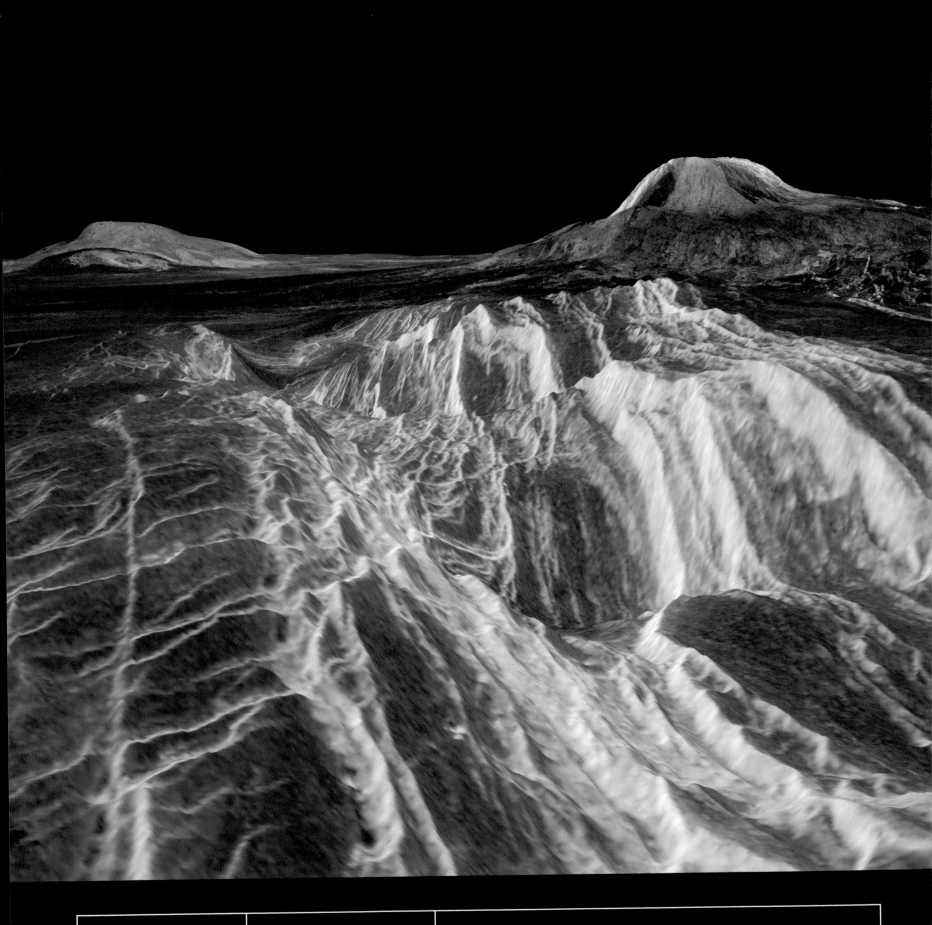

Venus

Sif and Gula Mons

million km
from the Sun

Volcanic mountains

While on Earth volcanism triggered by tectonic plates continually allows heat to escape from the planet's core, it's thought that heat inside Venus has no such escape, so every billion years or so the planet 'boils over' in a period of intense volcanism that wipes away much of what went before. However, no one knows for sure if volcanoes such as Sif and Gula Mons might still be active.

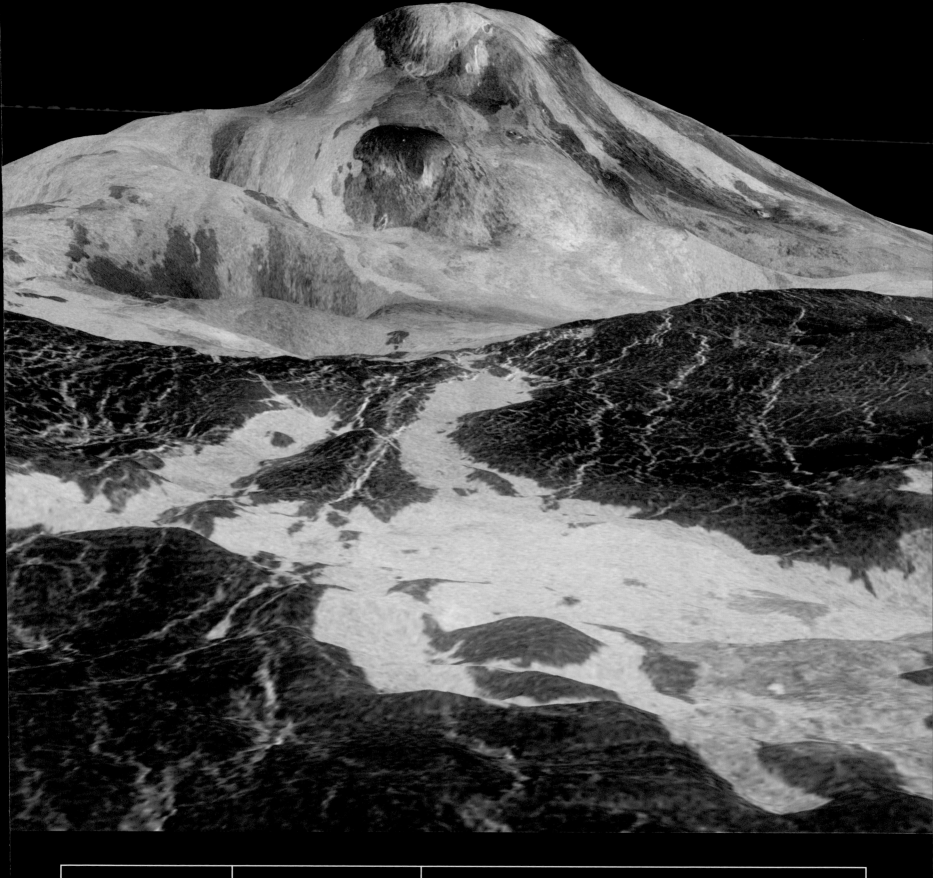

Venus

Maat Mons

Volcanic mountain

This image of Maat Mons exaggerates the volcano's height by a factor of ten. In reality, the peak is an impressive 5km (3 miles) above the surrounding plain, but its vast extent, spreading over hundreds of kilometres, makes it a gentle rise rather than a dramatic peak. This image uses Magellan data to portray slope and roughness as well as height, but the colour is an educated guess.

108.2

million km
from the Sun

Venus

Sapas Mons

Volcanic mountain

This Magellan view of Sapas Mons has been coloured to emphasise differences in the roughness of terrain. Such processing reveals a number of different lava flows forming layers around the volcano. The dark spots near the peak are flat-topped mesas, rather than calderas – Sapas Mons seems to have erupted most of its lava through vents around its slopes.

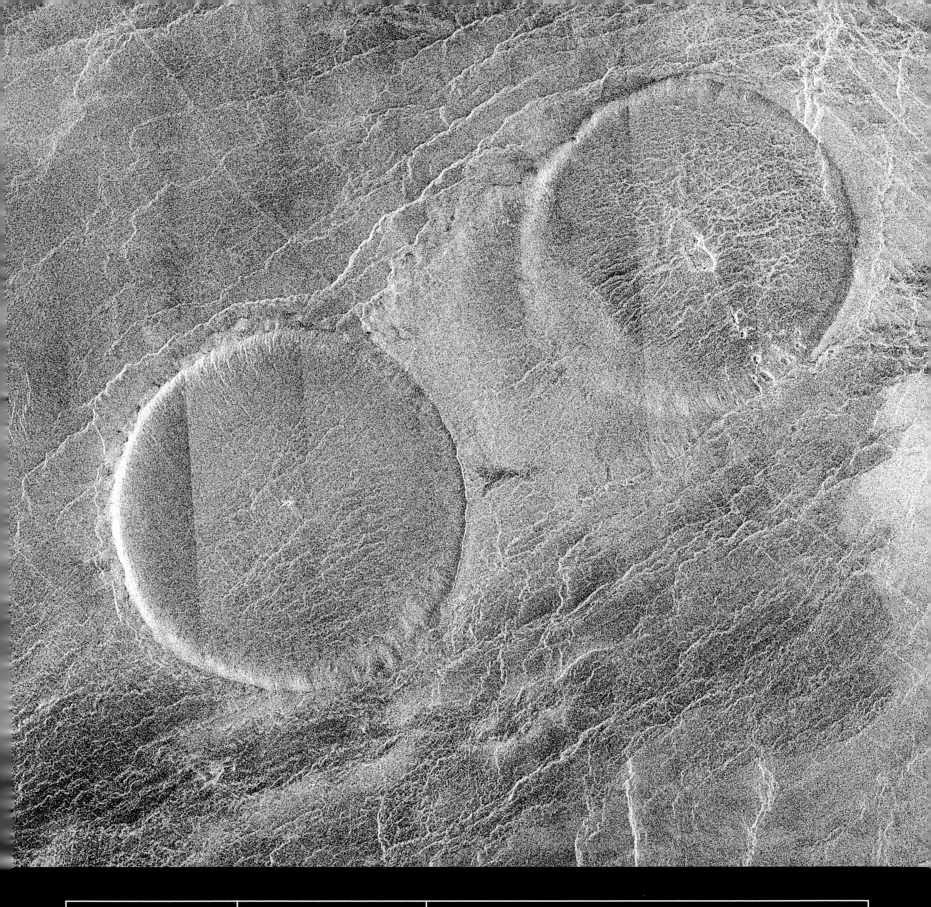

108.2

million km
from the Sun

Venus

Pancake domes

Volcanic outcrops

From overhead, it's hard to tell that these 'pancake domes' are raised features on the Venusian surface. These two domes in the Eistla Regio region are both 65km (40 miles) across, with flat tops 1 km (0.6 miles) above the surroundings. It's thought that they formed when particularly thick and viscous lava oozed through vents in the crust, setting before it could flow away from the fissure.

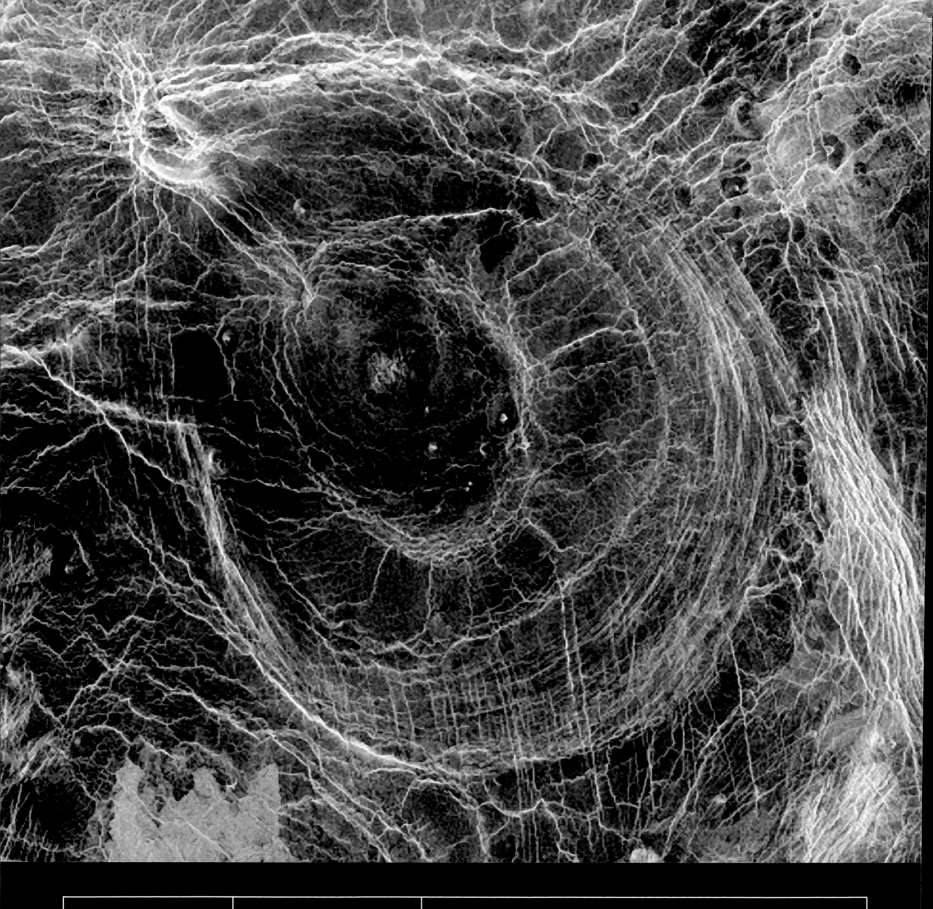

108.2

million km
from the Sun

Venus

Arachnoid

Volcanic depression

Arachnoids get their name from their appearance – they are sunken networks of concentric and radial cracks in the Venusian surface, forming a roughly circular, spider web-like pattern. They probably show areas where rising magma forced the crust to swell outwards. Eventually it cracked apart in these complex patterns, relieving the pressure and allowing the ground to subside.

Venus

Latona Corona and Dali Chasma

Tectonic features

Coronae are sunken areas surrounded by roughly concentric rings, and probably formed in a similar process of swelling and collapse to the arachnoids. Latona Corona, to the left of this image, is 1,000 km (620 miles) across. Dali Chasma, the trench to the right, is around 3 km (1.9 miles) deep, and is thought to be a fault line created during the planet's false-start at plate tectonics.

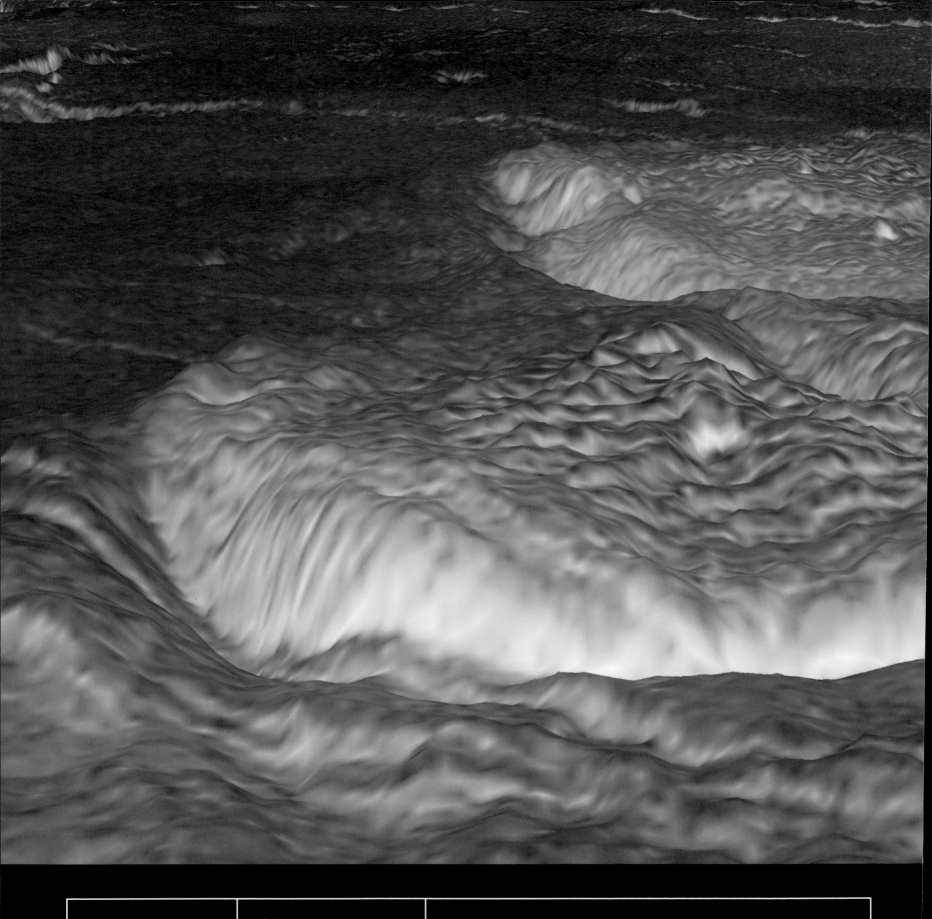

108.2

million km
from the Sun

Venus

Pancake domes

Volcanic outcrops

This three-dimensional view of a cluster of pancake domes in Alpha Regio exaggerates height differences to emphasise details. Each of these domes is around 25 km (15 miles) across, and about 750 m (2,500 ft) high. Cracks in the top of the domes probably formed as the lava source beneath the domes disappeared and the domes themselves subsided.

108.2

million km
from the Sun

Venus

Daniloca, Howe and Aglaonice

Impact craters

Craters on Venus are few in number, but generally quite large – the dense atmosphere shields the planet from smaller incoming meteorites, so that only the largest objects reach the surface. Even then, they may break up into a number of smaller fragments, as probably happened above Lavinia Planitia where this trio of craters formed millions of years ago.

108.2

million km
from the Sun

Venus

Dickinson crater

Impact crater

As material is blasted out from the site of a Venusian impact, the viscous atmosphere prevents it from travelling far, confining it to thick 'lobes' around the crater rim. The lobes tend to splash away from the incoming meteorite, revealing the trajectory of the original impact. In this crater, the impact also seems to have triggered volcanic eruptions, partially flooding the crater floor.

Earth – the homeworld

The arrival of remote sensing satellites and spaceprobes has transformed mankind's vision of its home planet almost as much as it has changed our understanding of our planetary neighbours. Studies of Earth made from orbital vantage points both near and far have the benefit of providing the 'bigger picture' that Earthbound scientists, restricted to recording data at localised sites, cannot get in any other way. Meanwhile, the transformation of the other planets and moons from distant blurry disks into complex changing worlds has provided us with some points of comparison, by which we can judge the features that make Earth special.

The most obvious of these is the presence of liquid water in huge amounts on Earth's surface. While H_2O is found in abundance throughout the Solar System, it is generally in the form of frozen 'water ice' – only on Earth does it thrive as a liquid and a vapour in the atmosphere. This is largely because of Earth's location at a 'sweet spot' in the Solar System – close enough to the Sun for the surface to be reasonably warm, but not so close that water is boiled off into space. The Earth's relatively thick atmosphere also plays an important role here, forming an insulating blanket that moderates temperatures on the surface, and exerting enough pressure to slow down the evaporation of large bodies of water. In turn, the atmosphere is only preserved because of Earth's relatively large size and mass – if Earth's gravity were weaker, then molecules from the atmosphere would escape more easily into space and the atmosphere would rapidly become rarefied.

Another major factor that distinguishes Earth from the other worlds of the solar system is its 'tectonics'. Instead of forming a solid shell of rock, Earth's crust is uniquely, broken into numerous tectonic plates some the size of continents. The plates drift around on top of the Earth's churning, semi-molten mantle, pulling apart from one another in some places, grinding past each other in others, and sometimes colliding head-on. The interaction between plates destroys and renews the Earth's crust, creates mountains, and triggers volcanoes. Although some other worlds show signs of ancient, stalled tectonics, the Earth's active system is unique. It is driven by heat from inside the planet – heat partially left over from the Earth's formation, and partly released anew by the decay of radio-active elements. Both factors will naturally have more effect for larger planets, which explains why they are strongest on Earth. The abundance of water also seems to play an important role in 'lubricating' the Earth's tectonics.

The third great difference between Earth and any other known world is the presence of life. This too may be traced, at least partially, back to the existence of surface water. In its liquid form, water acts as an ideal 'solvent' in which many of the organic (carbon-based) chemical building blocks of life can dissolve and undergo chemical reactions with each other. On a 'dry' world, such chemicals might still form or be carried to the surface (for example, by comet impacts), but without a solvent to aid their mobility, they would find it much harder to come together and react.

The camera-laden spy satellites of the Cold War were the ancestors of later 'remote sensing' satellites that use a wide range of techniques to glean more information about the world below them. Thanks to satellites, we can now monitor global weather patterns and study the impacts of climate change. We can map the ocean floors and tectonic boundaries, and find geological features ranging from underground water courses to lost cities. We can even study the distribution of plant life across our planet, and detect the spread of wildfires and plant diseases. The rise of the satellite age has transformed every aspect of our knowledge about Earth.

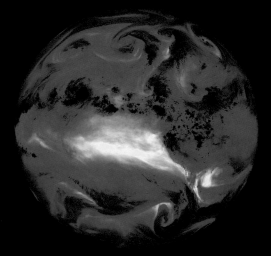

Earth's atmosphere is a huge machine for transferring heat between hotter and colder regions. A series of vast convection cells moves warm air away from the equator and towards the poles, while the coriolis forces caused by Earth's rotation creates patterns of prevailing winds. Infrared satellite maps of water vapour help to reveal the complex circulation patterns that result.

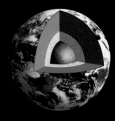

Earth's crust is dominated by silicate rocks, a few tens of kilometres deep. Beneath this lies the mantle of solid but mobile rock, then a core of iron and nickel, molten on the outside but solid in the middle.

Distance from sun	Orbital period	Orbital eccentricity	Diameter	Surface gravity	Rotation period	Axial tilt	Natural satellites
149.6	365.25	0.016	12,756	1	23.93	23.45	1
million km	Earth days		km	g	hours	degrees	

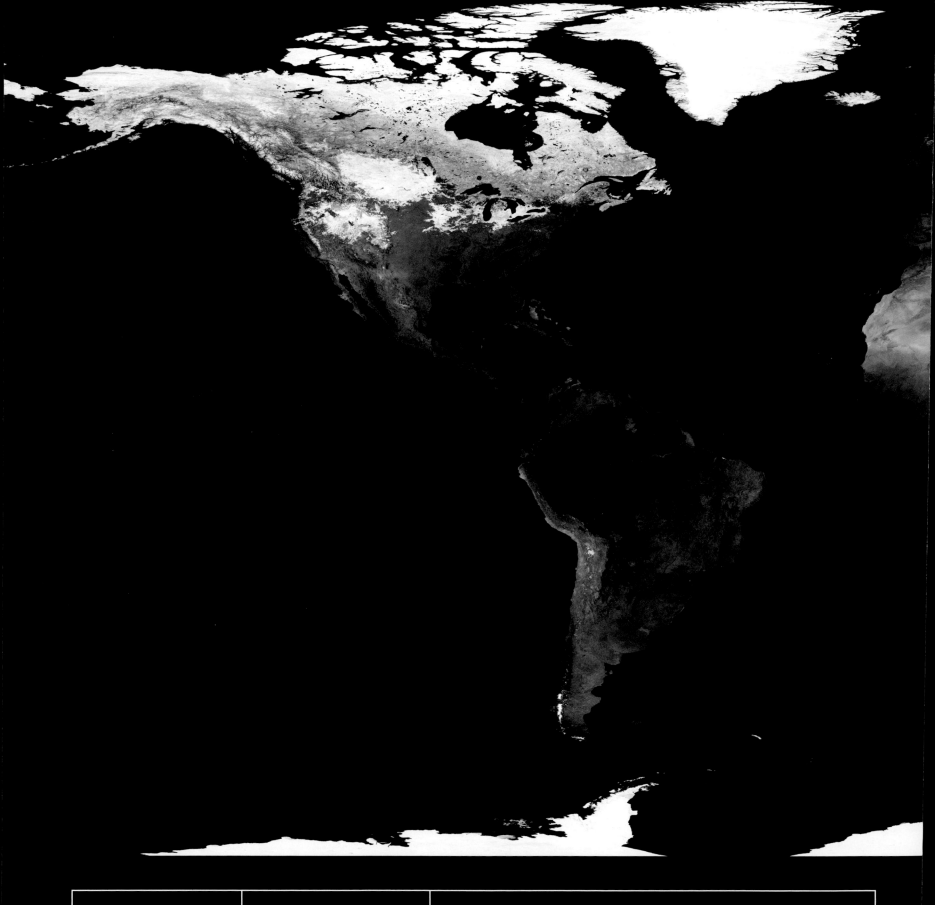

149.6

million km
from the Sun

Earth

Earth without clouds

Terrestrial planet

NASA photomosaics of the entire planet Earth reveal how our planet, uniquely in the Solar System, is dominated by water. It's not surprising that the team behind these images nicknamed our planet 'The Blue Marble'. Note that in this map, the higher latitudes are 'stretched' horizontally compared to the equator, in order to create a rectangular projection.

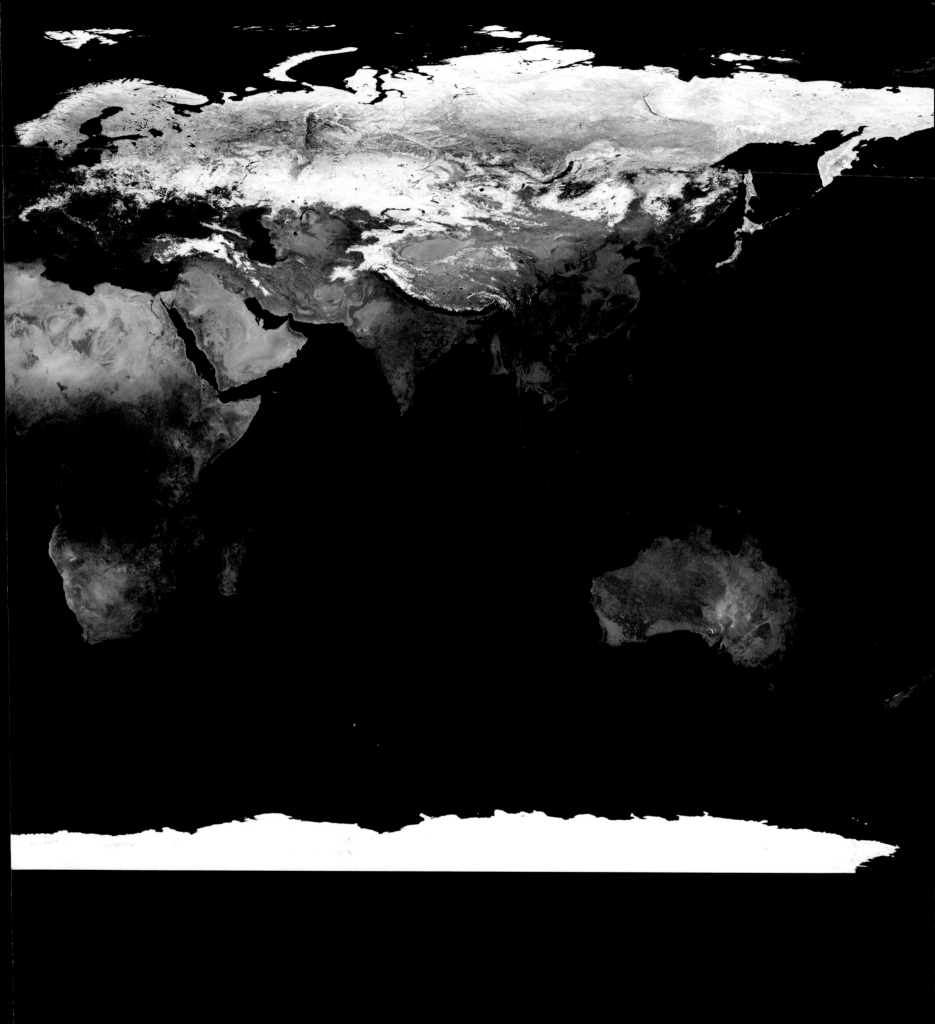

million km
from the Sun

Earth

Himalaya Mountains

Tectonic mountain chain

As tectonic drift has forced the Indian subcontinent into mainland Asia over the past 50 million years, the northern edge of the Indian plate has crumpled and been forced upwards to create the Himalayas. In order to support the mountain chain and the Tibetan Plateau (at the centre of this satellite image), the Earth's crust in this region protrudes downward like the unseen portion of an iceberg.

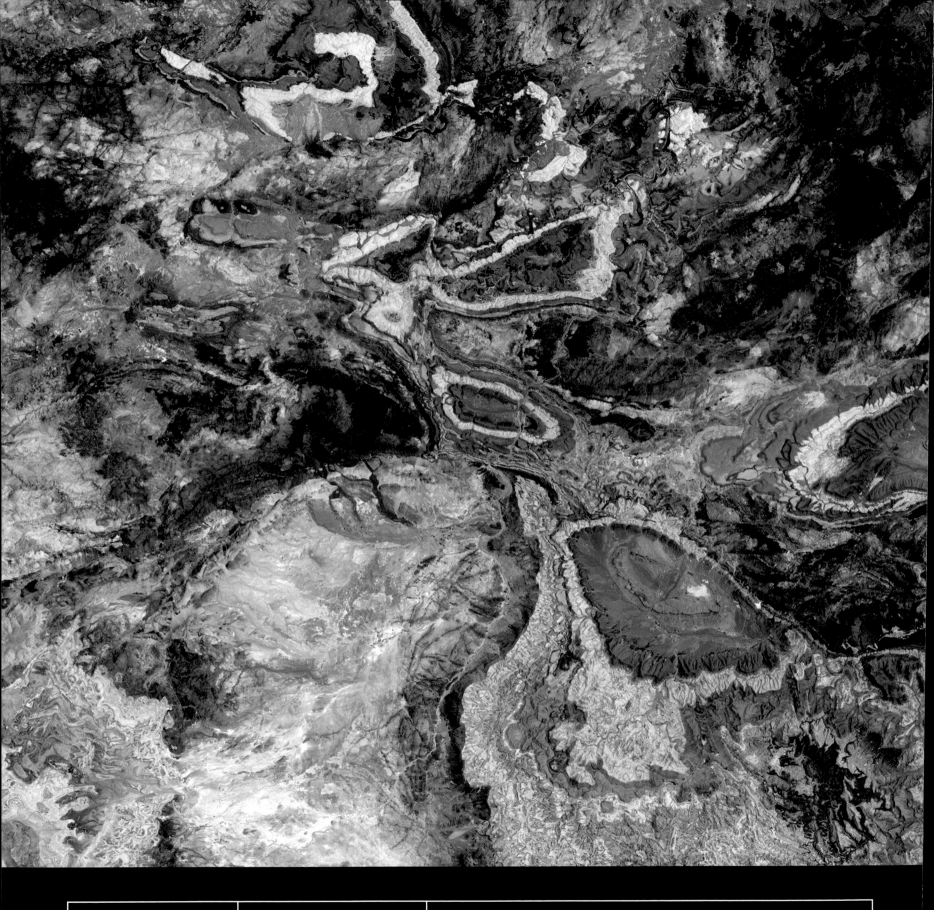

149.6

million km
from the Sun

Earth

Atlas Mountains

Tectonic mountain chain

Swirling colours indicate the presence of different minerals in this image from the Landsat 7 satellite. The Atlas range of northern Africa has formed in the past 60 million years as the African and European continents collided, but this region, the Anti-Atlas, contains remains of ancient mountains that have been through the entire process of uplift and gradual erosion once before.

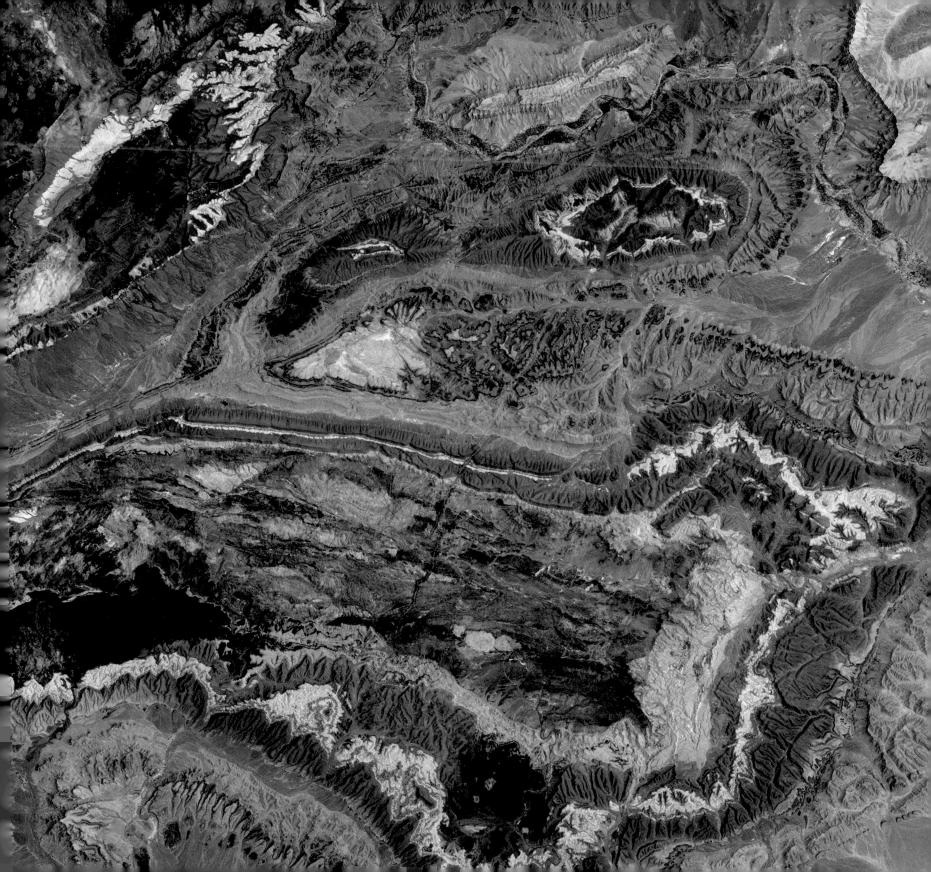

149.6

million km
from the Sun

Earth

Death Valley

Tectonic fault

California's Death Valley is one of the hottest, driest places on Earth. Vegetation clings to a series of parallel ridges, while the valleys between them are parched (the blue features are dried-up salt pans). The valley and its surroundings mark a region where the North American plate has stretched, fracturing into a series of parallel blocks that then 'toppled over' to create a zig-zag profile.

149.6

million km
from the Sun

Earth

Pampas Luxsar

Volcanic mountain chain

This vibrant image of the Andes in northern Chile reveals a landscape shaped by volcanic activity, with vegetation (red) clinging to the sides of volcano cones. The Andes show another way to build a mountain range – as the Nazca plate of the South Pacific is pushed below the South American plate and melts in the Earth's upper mantle, the energy released has created a long chain of volcanoes.

149.6

million km
from the Sun

Earth

Amazon rainforest

River basin

Life thrives in every conceivable niche on our planet, but nowhere is it more abundant than in the rainforests that girdle the Earth's tropical regions, where rainfalls usually exceed 2,000 mm (80 in) every year. The Amazon Basin (shown here) occupies about the same land area as the contiguous United States, and is home to at least 14,000 unique plant species, and countless animals.

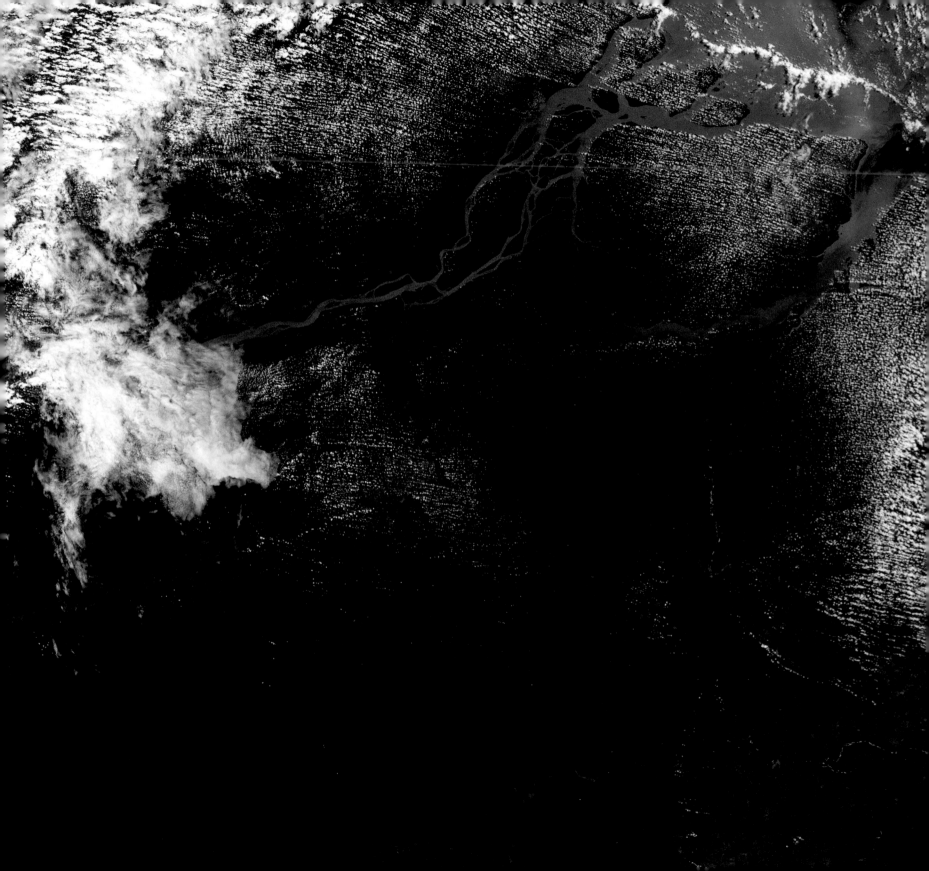

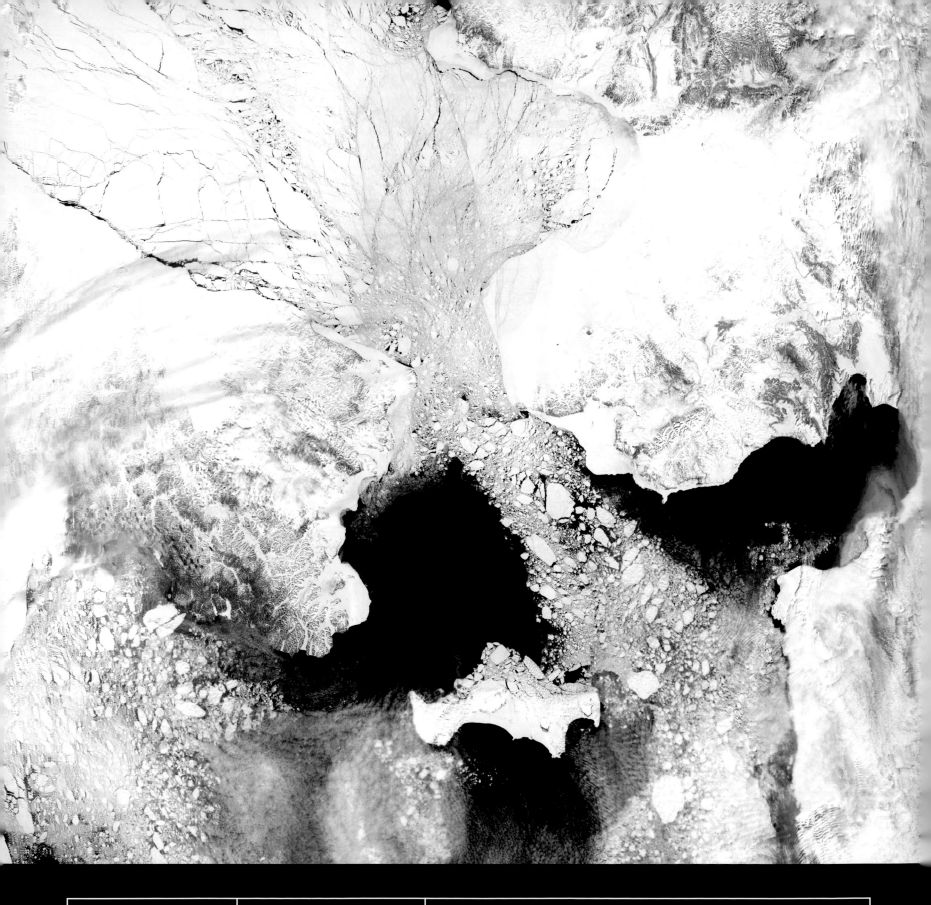

149.6

million km
from the Sun

Earth

Bering Strait

Water feature

Water exists in three phases across the surface of the Earth – as liquid, vapour, and ice. Here, the great ice sheet floating on the Arctic Ocean breaks into a multitude of small icebergs as it streams through the Bering Strait between Siberia and Alaska. The ice sheet grows and retreats with the changing of the seasons, which is most extreme in the Arctic and Antarctic circles.

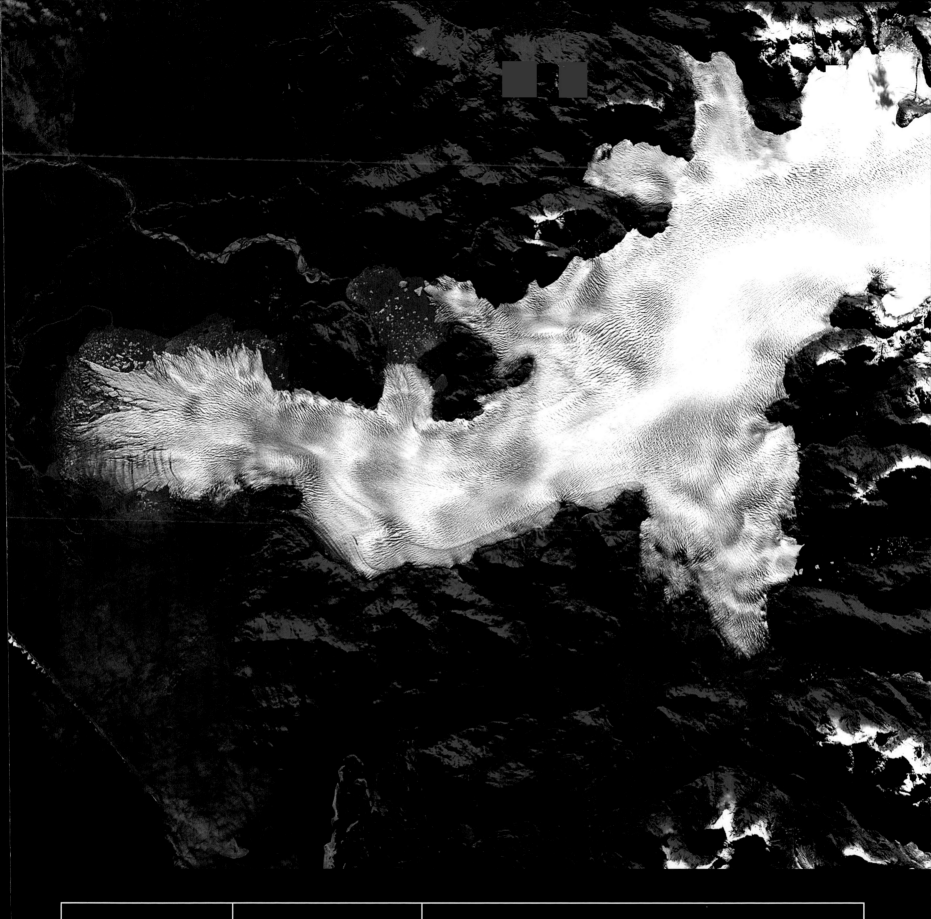

149.6

million km
from the Sun

Earth

Patagonian Ice Field

Water feature

Away from the poles, permanent ice exists only at high altitudes – anywhere that snowfall in winter does not completely melt in summer. The Patagonian Ice Field runs through the southern Andes, and is the third largest mass of ice on land after Antarctica and Greenland. But it is just a small remnant of a vast ice sheet that spread across South America during the Earth's last ice age.

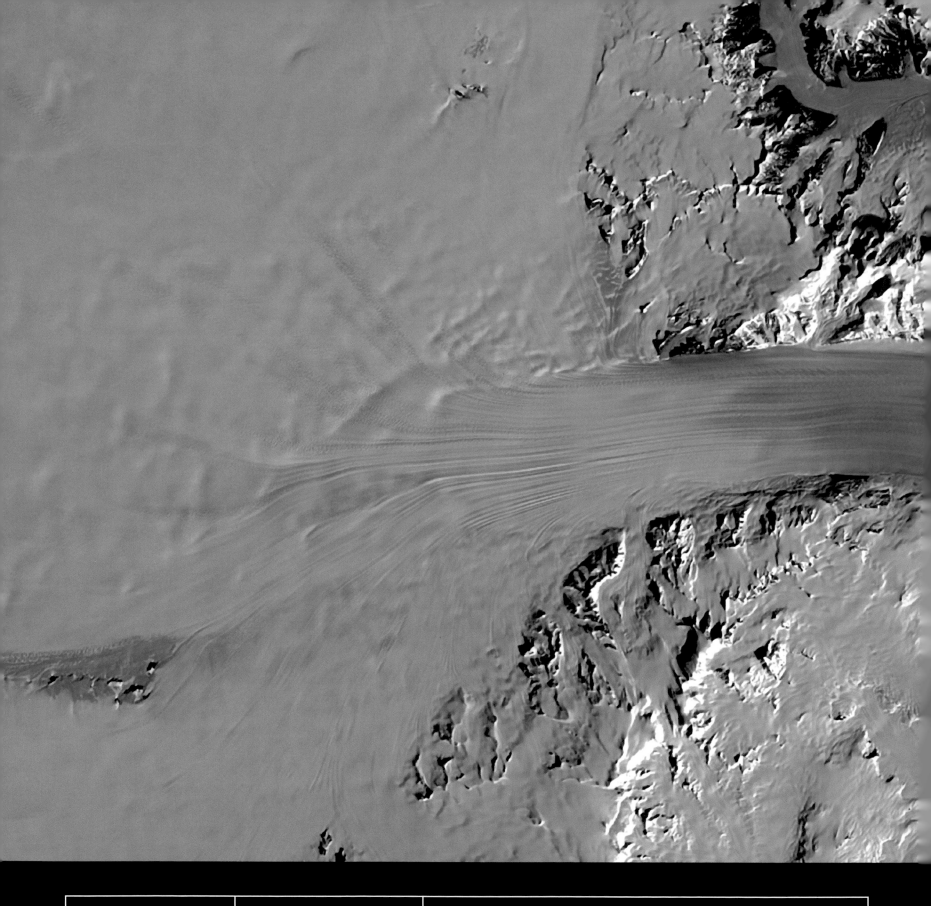

149.6

million km
from the Sun

Earth

Byrd Glacier

Water feature

Ice on Earth is not static, but flows slowly under its own weight. Antarctica's Byrd Glacier, seen here where it carves its way through the Transantarctic mountain range, is a spectacular river of ice 24 km (15 miles) wide and 160 km (100 miles) long, flowing at a speed of around 0.8 km (0.5 miles) per year from the raised polar plateau onto the floating mass of ice called the Ross Ice Shelf.

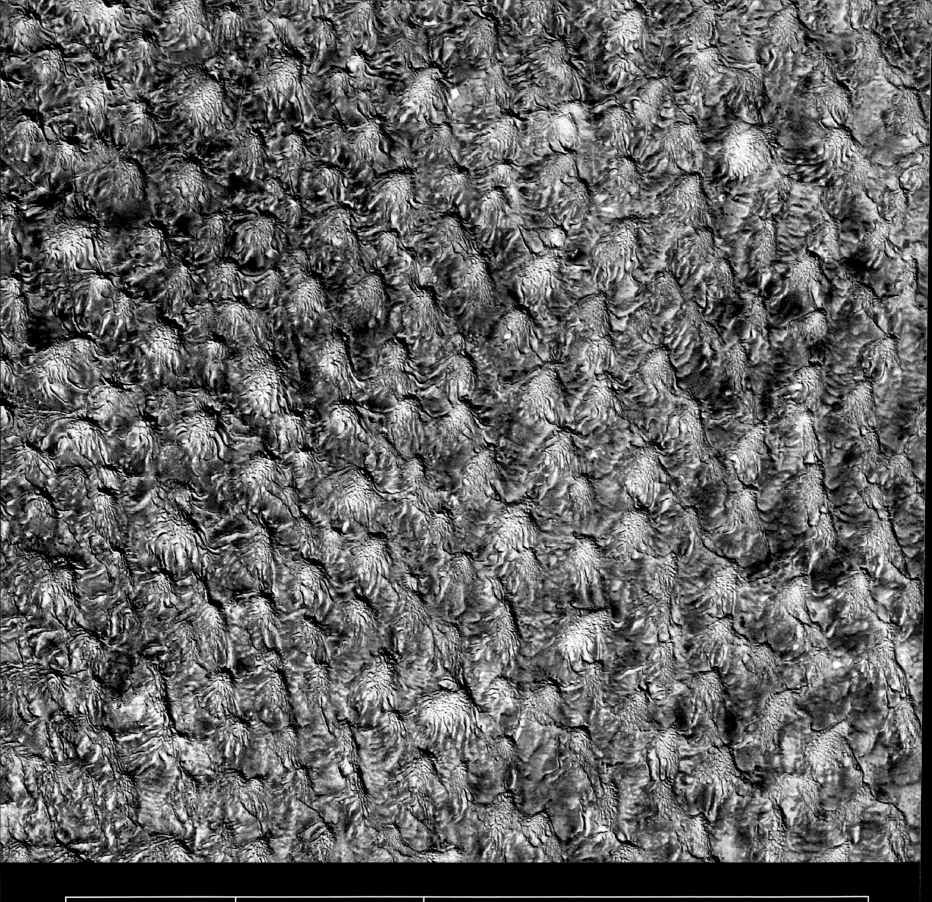

149.6

million km
from the Sun

Earth

Grand Erg Oriental

Desert

What appears at first to be an abstract image is in fact a huge field of star dunes, created by shifting winds in the Grand Erg Oriental, a region of the Sahara Desert in southern Tunisia. The Sahara is the largest of all Earth's deserts, with an area greater than the continental United States, but deserts occur wherever local conditions lead to very low annual rainfall.

million km
from the Sun

Earth

Rub al Khali

Desert

Barchan dunes in the Arabian Peninsula have formed from the slow but steady action of winds from a single direction over many centuries. Sand accumulates in deserts as erosion and weathering wear away soft rocks. The unique colours of this image show two distinct layers – a salt-rich blue silt has been overlaid with drier yellow sand, but then exposed again by the dune-forming winds.

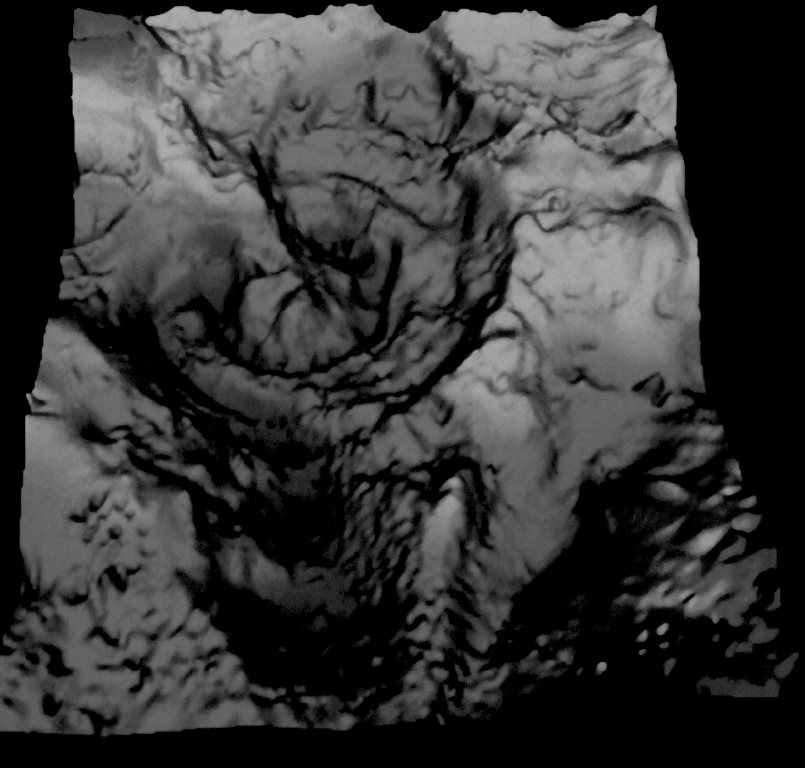

149.6

million km
from the Sun

Earth

Chicxulub Crater

Impact crater

Earth's atmosphere protects it from incoming space debris, but occasional major impacts are inevitable. 65 million years ago, a 10-km (6-mile) meteorite struck the Gulf of Mexico, forming a crater 240 km (150 miles) across that is widely linked to the extinction of the dinosaurs. Today, the crater is buried beneath layers of sediment, its structure only revealed by magnetic and gravity surveys.

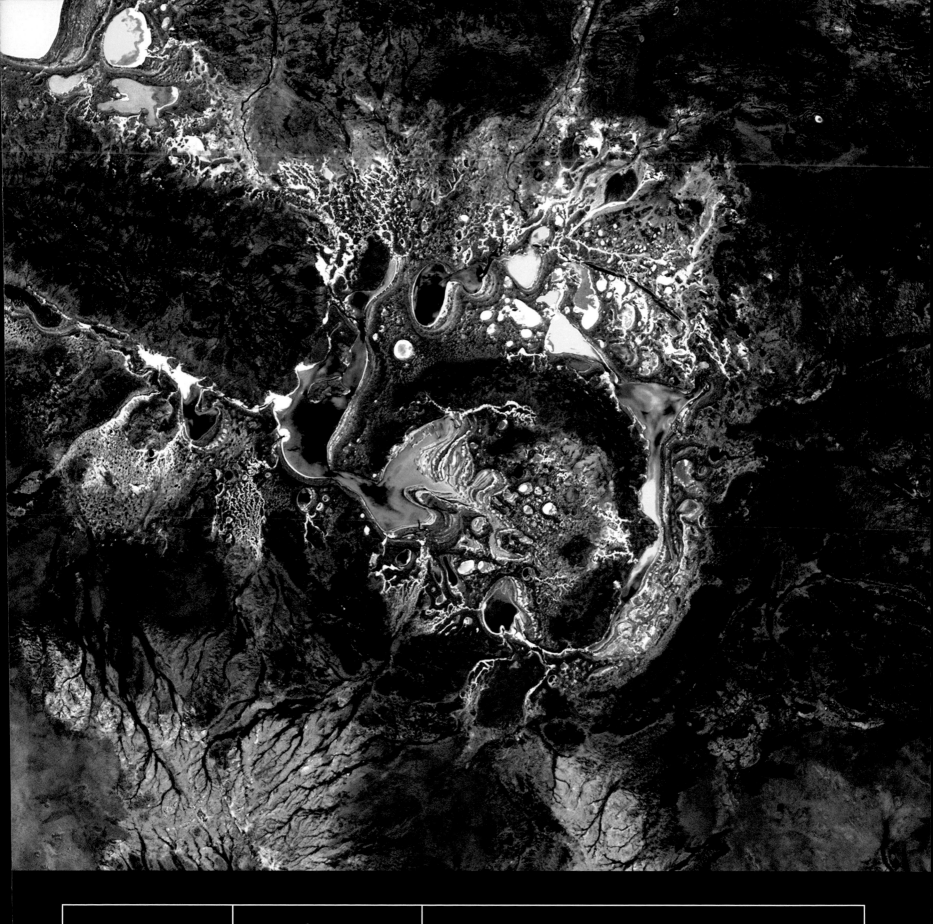

million km
from the Sun

Earth

Shoemaker Crater

Impact crater

Looking like an artist's paint-spattered palette, this satellite image shows the Shoemaker crater, an ancient but well-preserved impact in the desert of Western Australia. Formed 1.7 billion years ago, the crater is about 30 km (18 miles) in diameter, though aeons of weathering has disrupted its shape. Today, lakes in and around Shoemaker have produced salt deposits in a variety of bright colours.

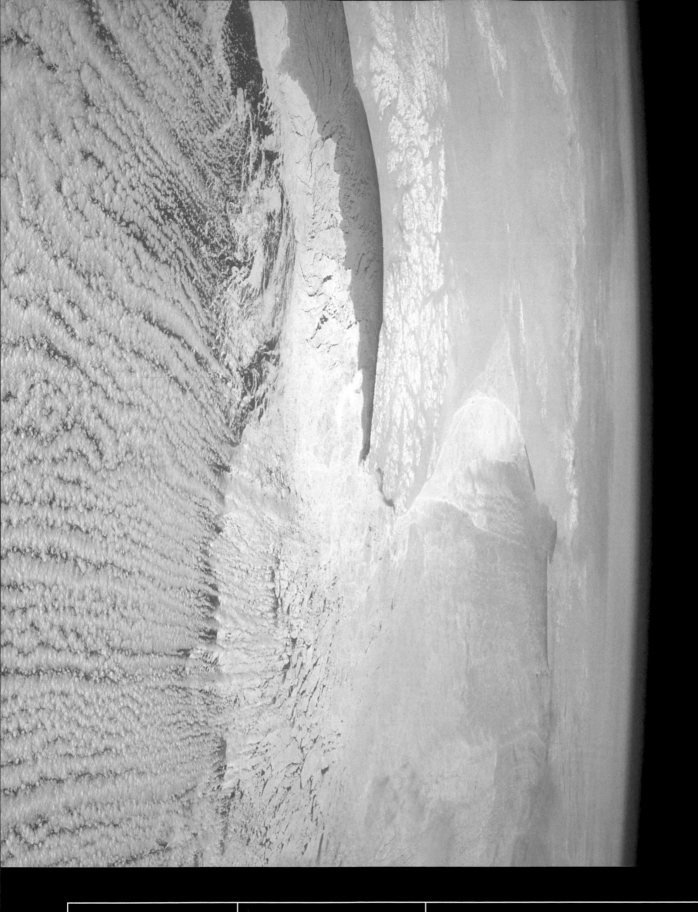

149.6

million km
from the Sun

Earth

Cloud streets

Atmospheric feature

Earth's atmosphere is a protective halo of gas just 190 km (118 miles) thick. Yet this thin mix of nitrogen, oxygen and other trace gases has created the hospitable planet we enjoy today, protecting the planet from extremes of temperature and shielding it from high-energy solar radiation, particles from the solar wind, and the vast majority of stray space debris.

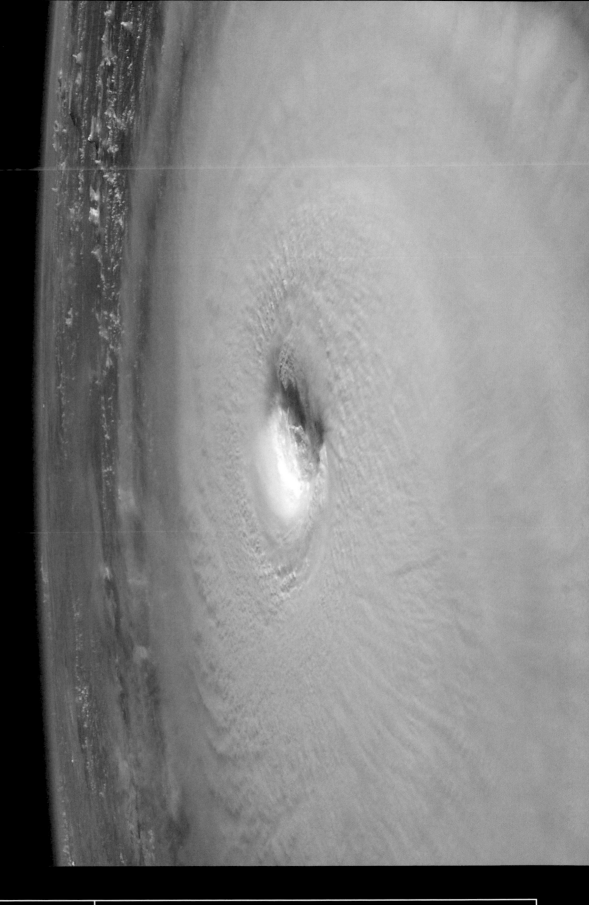

million km
from the Sun

Earth

Hurricane Isabel

Atmospheric feature

Earth's atmosphere boils with energy as heat flows from equator to poles and sun-warmed pockets of water vapour rise, cool, condense and fall. One of most impressive manifestations of all this energy trading is the hurricane, a heat engine that feeds off warm seas, and is capable of producing an hourly energy output equivalent to a 30-megaton nuclear detonation.

149.6

million km
from the Sun

Earth

Aurorae from space

Atmospheric feature

Earth is sheathed in a magnetic field that loops from pole to pole, creating a protective barrier from high-energy particles from the Sun. Swept up in this field, charged particles ricochet back and forth, filtering down to Earth at the north and south magnetic poles. Here they interact with molecules in the upper atmosphere to produce glowing curtains of light – the aurorae.

Destination Moon

Earth's only natural satellite was the inevitable first target for interplanetary exploration. As early as 1959, a series of Soviet spaceprobes had flown past it, smashed into it, and swung behind it to send back pictures of the elusive far side. (The Moon, like most planetary satellites, has had its rotation slowed by tidal forces so that it now spins on its axis in the same time it takes to circle the Earth, and therefore keeps one face permanently towards our planet, and the other perpetually turned away.)

The armada of visiting spaceprobes continued throughout the 1960s, as the Moon became the ultimate goal in the final phase of the space race between the United States and the Soviet Union. US efforts included the Lunar Orbiter series of spacecraft, which as the name suggests slipped into orbit around the Moon and compiled an extensive photographic atlas of its surface. The Ranger spacecraft, meanwhile, were sent hurtling into the lunar landscape, taking photographs all the way to impact. It was these craft that first revealed the extent of lunar cratering, and also its cause – until this time, many astronomers had believed the craters were volcanic in nature and would therefore have a lower size limit. Instead, they turned out to continue down to microscopic level, indicating that they had to be formed by meteorite impacts of varying sizes. Finally, the Surveyor missions were a series of craft that made soft landings on the surface, sending back pictures and scientific data. Importantly, they also proved the feasibility of landing a manned vehicle on the Moon, for until the Surveyors successfully touched down, the consistency of the lunar soil or 'regolith' was unknown, and some astronomers had fretted that it was no more solid than talcum powder, into which any substantial spacecraft would sink without trace.

All these robot missions, of course, only served to pave the way for the ambitious Apollo program, which saw the first landing on the Moon on 20 July 1969. By this time, the Soviet Union's own plans for a manned lunar mission had fallen by the wayside, but US astronauts completed five further landings, with a total of twelve men eventually walking on the Moon before the end of 1972.

The Apollo landings targeted a variety of different landscapes across the near side of the Moon, recording the detailed geology and collecting rock samples. Elsewhere in the Solar System, we can only guess the relative ages of different surfaces, looking for clues where one 'unit' of the terrain overlaps its neighbour, or simply counting the number of impact craters that have accumulated on different landscapes in the time since they formed. Rocks from the Moon, however, can be dated directly using the same 'radiometric' techniques archaeologists and palaeontologists use on Earth – the rate at which radioactive elements decay indicates how much time has passed since the rock formed. Most of the cratering that peppers the Moon seems to have happened in the first 500 million years of its existence, peaking around 3.9 billon years ago, and culminating in a series of impacts from larger objects that formed huge impact basins and wiped the surface beneath them clear. Shortly after these impacts, many of the basin floors cracked open, allowing molten lava to well up from within the Moon. As it set solid, it formed the smooth, undulating plains known as the lunar seas or 'maria'. Since then, it seems, cratering has continued at a much slower rate, leaving the maria still unmarked to this day.

The Apollo program was deliberately designed to get American astronauts to the Moon in a hurry – not to pave the way for a long stay. Throughout the 1970s and 1980s, visionary plans for moonbases were ignored in favour of space travel closer to home. It was not until the 1990s that interest in the Moon was reignited, thanks to the Clementine and Lunar Prospector probes. Clementine was the first satellite to use radar on the moon, building up the first detailed map of lunar surface altitudes. Lunar Prospector, meanwhile, used familiar Earth satellite remote sensing techniques to identify minerals in the lunar surface. Both probes found tantalising evidence for water lurking in the permanently shadowed craters at the lunar poles, perhaps dumped there during comet impacts early in the Solar System's history.

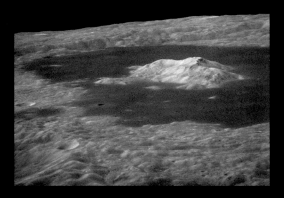

The Apollo 12 mission landed in the lunar Ocean of Storms, close to an unmanned lander called Surveyor 3. The astronauts visited and photographed it to see how it had fared during its time on the Moon.

Top: While the lunar nearside has many broad, dark seas, the far side is far more mountainous. The most substantial lava plain is on the floor of Tsiolkovskii, a medium-sized crater.

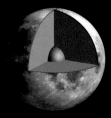

The Moon has a fairly simple internal structure, befitting a small world that cooled and solidified rapidly. Beneath a crust of volcanic rock, thicker on the far side than the near side, lies a deep mantle of solidified rock, perhaps with a small core of solid metal at the centre.

Distance from Earth	Orbital period	Orbital eccentricity	Diameter	Surface gravity	Rotation period	Axial tilt	Natural satellites
384,400	27.32	0.05	3,476	0.16	27.32	1.54	-
	Earth days		km	g	Earth days	degrees	

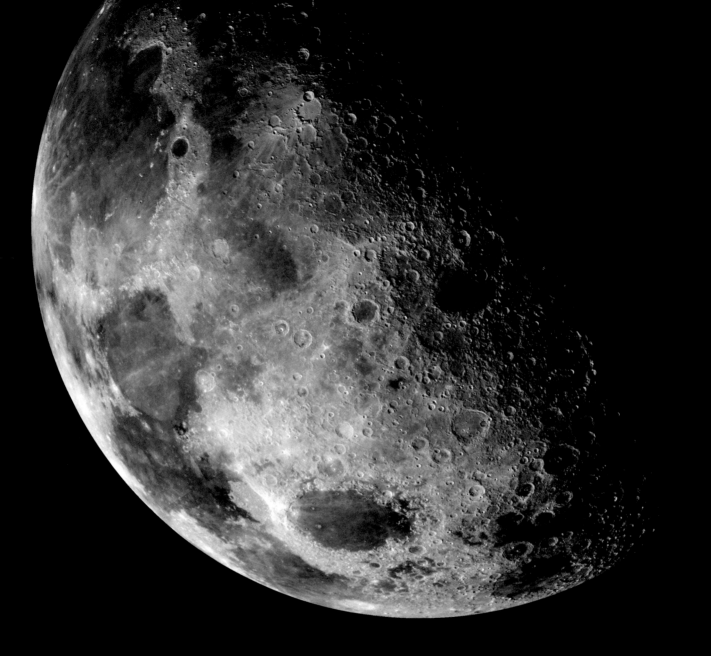

384.4

thousand km
from the Earth

The Moon

North polar region

Rocky natural satellite

In 1990, the Galileo probe flew past the Moon on its way to Jupiter, sending back this image from 425,000 km (262,000 miles) above the north pole. Familiar features of the lunar nearside can be clearly seen, including the compact dark sea of the Mare Crisium near the bottom edge, and the Sea of Tranquillity on the limb at the 'eight o'clock' position.

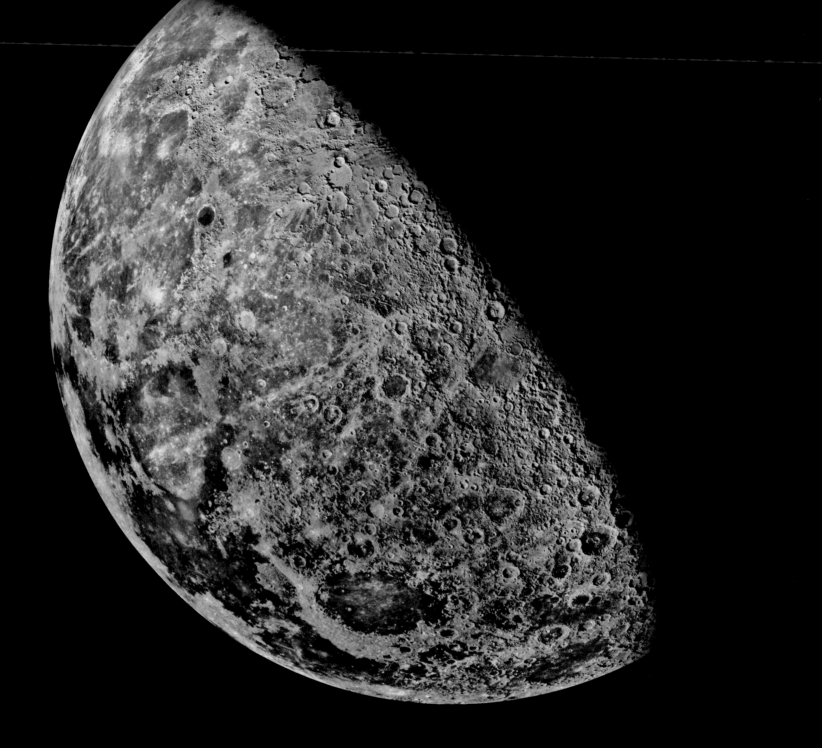

384.4

thousand km
from the Earth

The Moon

North polar mineral map

Rocky natural satellite

The Moon's colours are subtle but definite – different minerals tint rocks of varying ages and origins with different hues. Here the Galileo image opposite has been processed to highlight colour differences. The result shows how rock 'units' of different types and ages overlay one another. Red and green areas are ancient highlands, while blue and orange are maria or seas.

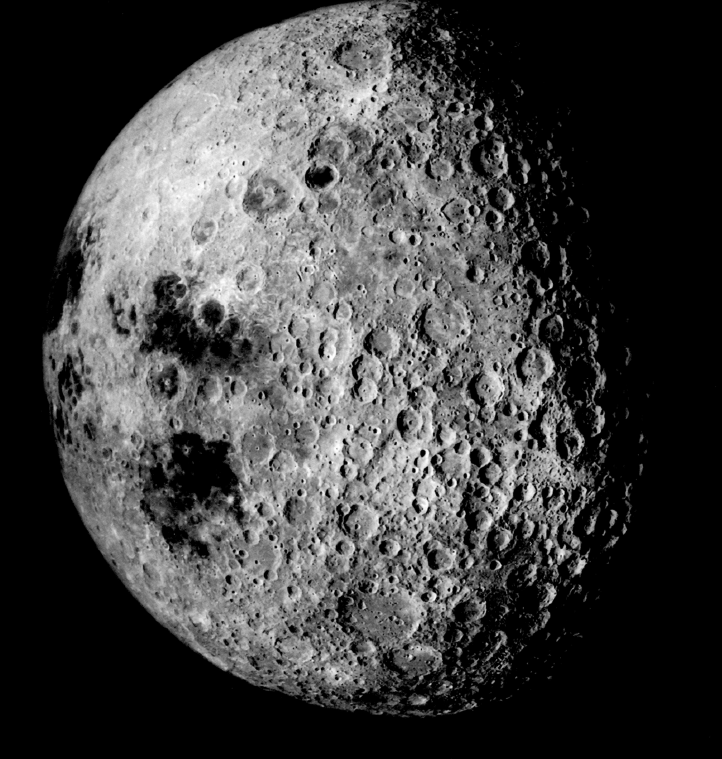

384.4

thousand km
from the Earth

The Moon

Lunar far side

Natural rocky satellite

The far side of the Moon is quite different from the near side, with only a few small seas darkening a landscape of bright highland material. There are still impact basins on this side of the Moon, but the effect of Earth's gravity meant that the erupting mare lava which flooded the nearside basins had a much harder time reaching the far side surface.

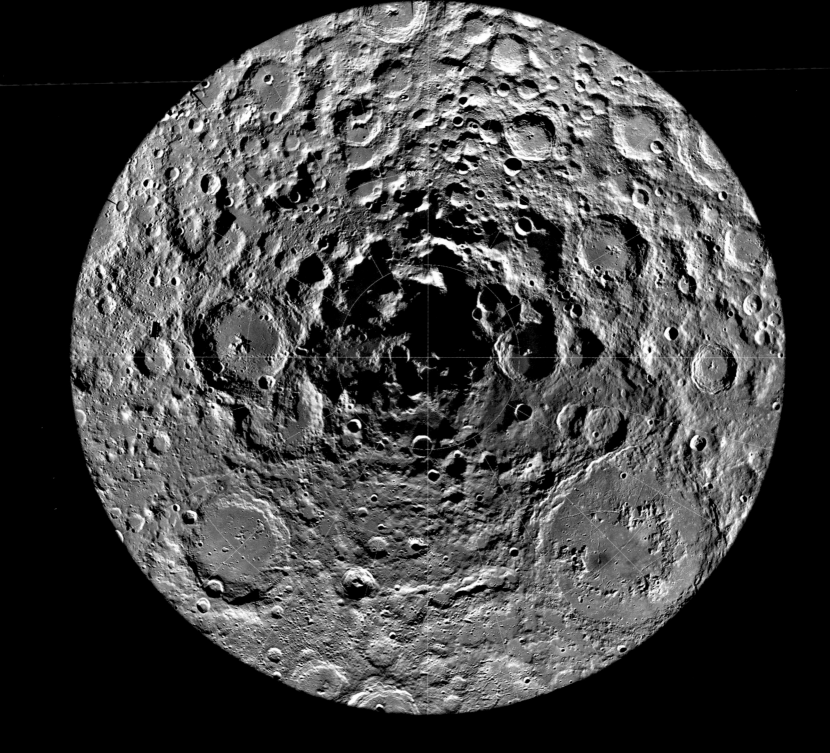

384.4

thousand km
from the Earth

The Moon

South Pole-Aitken Basin

Impact basin

Although its existence was suspected in the Apollo era, we had to wait for a new generation of lunar orbiter spaceprobes to confirm and photograph the Moon's biggest secret. Much of the far side is dominated by an enormous impact feature called the South Pole-Aitken Basin. At roughly 2,500 km (1,550 miles) across, the basin is larger than western Europe.

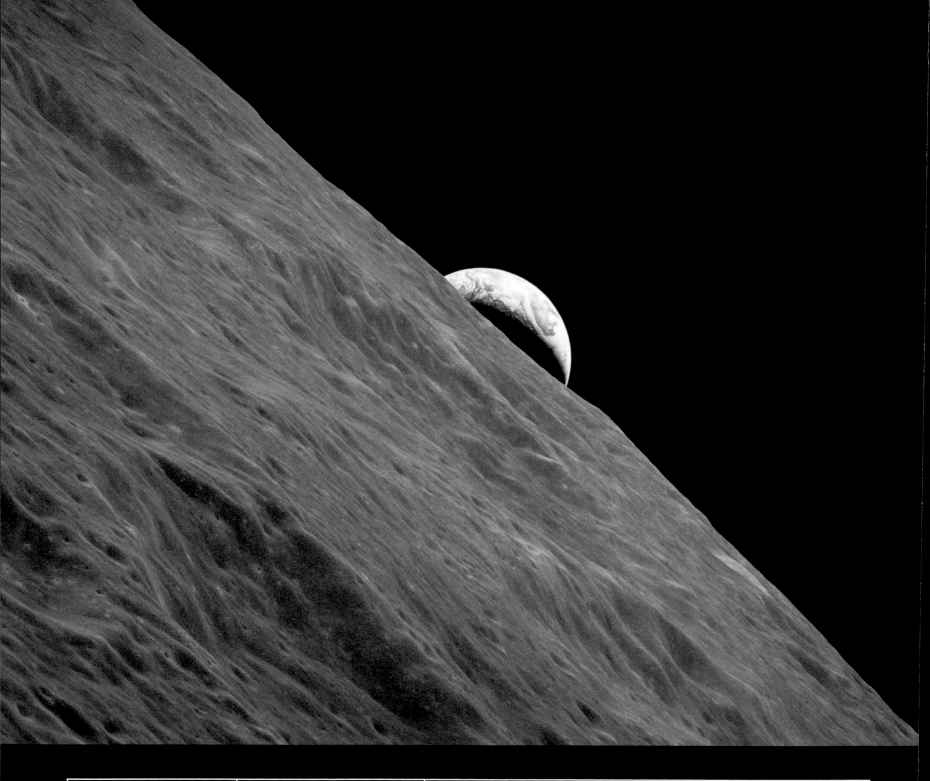

384.4

thousand km
from the Earth

The Moon

Earthrise over the Moon

Just as the Moon keeps the same face permanently towards Earth, so Earth barely wavers its position in lunar skies. The only way to see an 'Earthrise' such as those filmed by the Apollo astronauts is to circle the Moon, on the surface or in orbit. The Earth 'rises' along the transition from the far to the near side, marked here by the Mare Orientale.

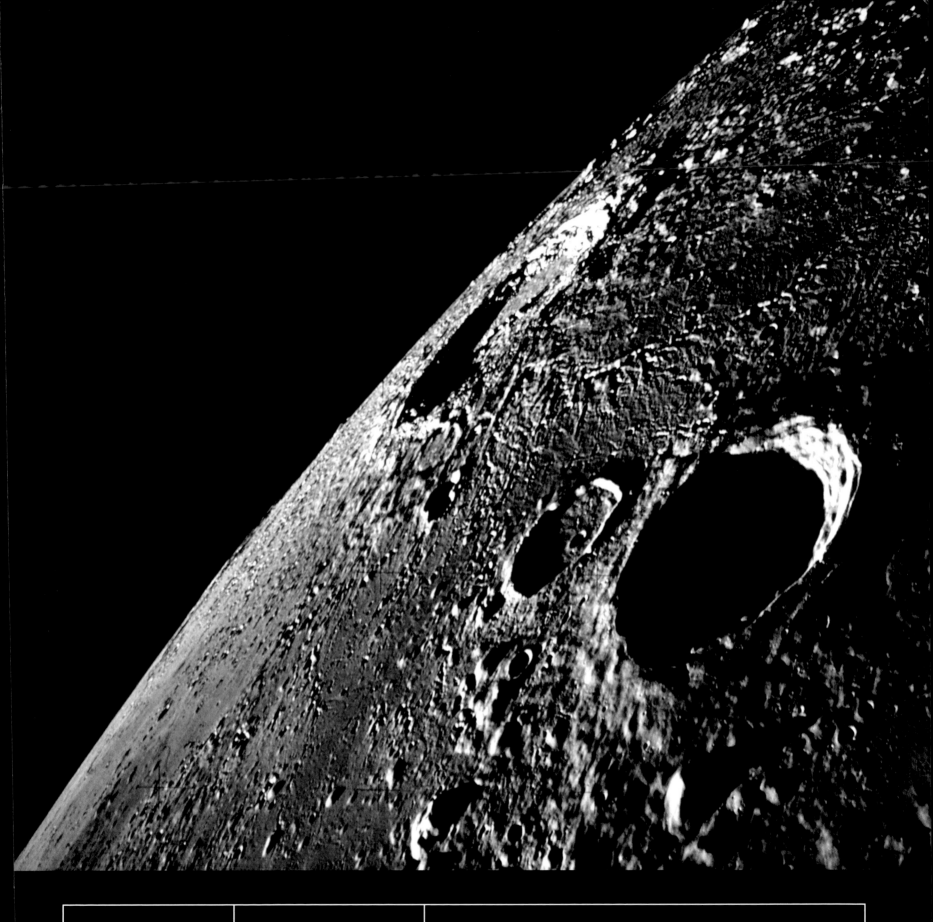

384.4

thousand km
from the Earth

The Moon

Copernicus and Eratosthenes

Impact craters

Copernicus, seen on the horizon in this image (with the smaller Eratosthenes in the foreground), is one of the youngest large craters on the Moon, formed about 900 million years ago. 91 km (57 miles) across and 3.7 km (2.3 miles) deep, it is surrounded by a bright blanket of ejecta – material thrown out of the crater by the impact. In places, the ejecta shot out in 'rays' up to 1,200 km (750 miles) long.

384.4

thousand km
from the Earth

The Moon

Sea of Tranquillity

Volcanic plain

This was the view from the Apollo 11 lunar lander Eagle just before it touched down on the Moon's surface on 20 July, 1969. The landing site in the Mare Tranquillitatis (Sea of Tranquillity) was a comparatively uninspiring rocky plain, but even then Eagle's pilot Buzz Aldrin had less than 30 seconds of fuel to spare when he found a suitably smooth area for landing.

 384.4

thousand km
from the Earth

The Moon

Astronaut footprint

Neil Armstrong took his first step onto the Moon at 02:56 GMT on 21 July, 1969, uttering the immortal words 'That's one small step for man… One giant leap for mankind'. A total of twelve astronauts have so far walked on the Moon, and, aside from chance meteorite strikes, their footprints should remain in pristine condition for hundreds of millions of years.

384.4

thousand km
from the Earth

The Moon

Shorty Crater

Impact crater

The lunar roving vehicle (LRV) carried aboard the later Apollo missions seems lost in the vast silent landscape of the Taurus-Littrow Valley in this panoramic mosaic photo. The crater to the right of the LRV was nicknamed 'Shorty' and is 110 m (456 ft) across. Orange patches in the soil around the LRV proved to be rich in zinc, iron oxide and titanium, probably with a volcanic origin.

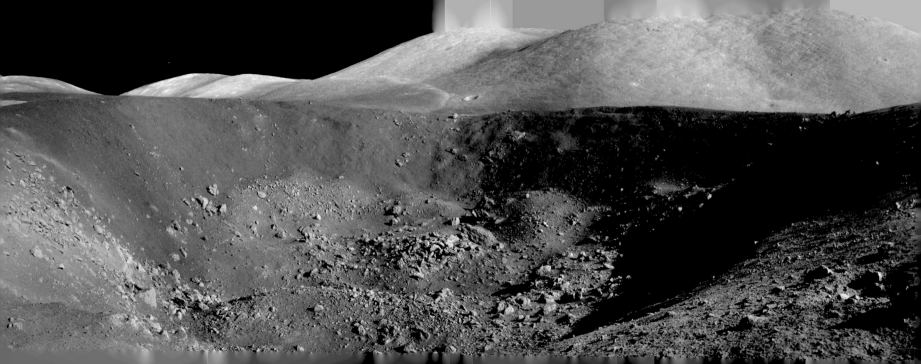

384.4

thousand km
from the Earth

The Moon

Taurus-Littrow

Lunar valley

Astronaut Harrison Schmidt of the Apollo 17 mission bounds across the eerie lunar landscape of Taurus-Littrow. With no atmospheric effects to aid perception, it is hard to appreciate the true distance of the hills behind him. Schmidt was the only qualified geologist to travel to the Moon, though the Apollo missions brought a total of 382 kg (824 lb) of moon rock back to Earth for study.

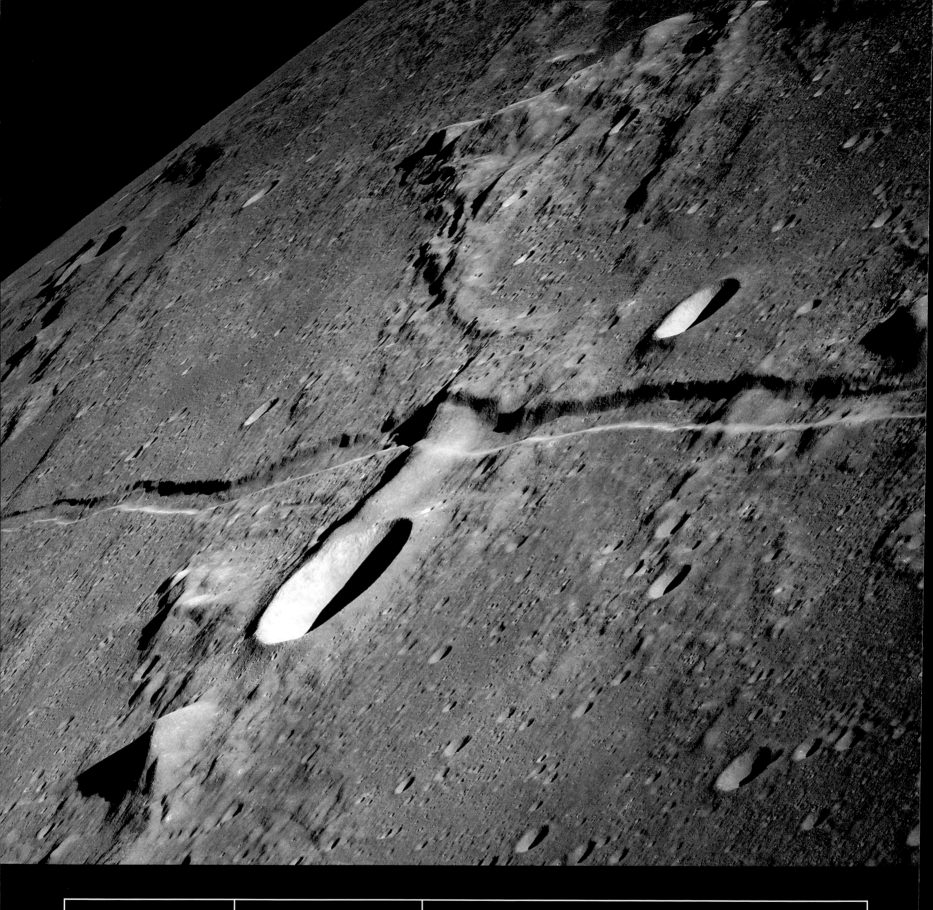

384.4

thousand km
from the Earth

The Moon

Aridaeus Rille

Volcanic lava tube

The lunar seas are by far the greatest volcanic monuments on the Moon, but there is also evidence of more localised activity. There are some small Earth-like volcanoes, and a number of 'rilles', valleys formed as lava tunnels beneath the surface collapsed. Aridaeus Rille, photographed here by Apollo 10, runs in a straight line for hundreds of kilometres.

384.4

thousand km
from the Earth

The Moon

Hadley Rille

Volcanic lava tube

The Apollo 15 mission landed close to one of the best-known rilles, a 120-km (75-mile) long winding channel that cuts its way through the Lunar Appenine mountains. Here, astronaut James Irwin stands alongside the Lunar Roving Vehicle on one side of the steep-sided Hadley Rille valley, which in this area is as much as 1.6 km (1 mile) across.

Red Mars

Perhaps more than for any other world, our perceptions of Mars have swung this way and that as successive generations have studied the Red Planet first with Earth-based telescopes, and later with each new wave of spaceprobes to slip into orbit or land on Mars.

Until the dawn of the Space Age, Mars, like Venus, was seen as a likely haven of potential life on the interplanetary doorstep. Even when the late-19th century furore about alleged 'canals' seen on the planet had died down (they were eventually proven to be nothing but an optical illusion), the odds still seemed to favour some form of life, if not intelligence, on the planet. The ice caps at each pole suggested that water was present here to some degree, and many people thought that the shifting dark patches on the surface were seasonal vegetation.

In 1964, after several false starts, the American space agency NASA finally sent a spaceprobe, Mariner 4, on a successful flyby mission to Mars. The handful of images it sent back were a crushing disappointment, for they revealed a landscape shaped mostly by heavy cratering that clearly hadn't changed for billions of years. The Martian atmosphere, meanwhile, proved to be a thin veil of carbon dioxide, incapable of warming the planet enough to sustain water on its surface, or protect any life struggling to evolve there from the ravages of solar radiation. To add insult to injury, it now seemed that the polar caps were largely made of frozen carbon dioxide ('dry ice') rather than water.

Two more flybys in the late 1960s seemed to confirm the terminal diagnosis – Mars was little more than a rusty version of our own Moon, a hostile and arid world. But by sheer coincidence, all these early Martian pioneers happened to fly over the same region of Mars – the area we now know as the southern highlands. By 1972, technology had improved enough to put the first spaceprobe in orbit around Mars, and take a look at the planet as a whole.

Mariner 9 arrived to find the surface of Mars entirely hidden by one of its periodic dust storms. As the atmosphere finally cleared after several weeks, a new side of Mars was revealed – a landscape of towering volcanoes, a truly gargantuan canyon system, and plentiful evidence that water had once shaped the landscape, ranging from slowly carved sinuous river valleys to the scars of catastrophic flash floods. Mariner was followed in the 1970s by the Viking orbiters and landers, which surveyed the planet in far more detail and sent back the first pictures and data from its surface. But although the proof of a wet Martian past was mounting, so was the evidence that modern Mars was a cold, arid desert. What, then, had become of the water? Some doubted it had ever really existed, and came up with theories to explain how the apparently eroded features could have formed in other ways.

Frustratingly, a series of probes sent to Mars in the 1980s and 1990s fell victim to a range of technical failures, and it was not until 1997 that robot explorers returned to the Red Planet in earnest. The Mars Pathfinder probe with its Sojourner rover proved that it was possible to drive a remote-controlled vehicle across the Martian surface to photograph and analyse whatever it came across. By 2004, it had been followed by two much more robust Mars Exploration Rovers, each targeted to land in areas that might once have been under Martian waters. Meanwhile, the Mars Global Surveyor orbited above the planet, returning the most detailed images yet, and was soon joined by other Martian satellites, including Mars Odyssey and the European Mars Express. These probes have supplied the new view of Mars presented in the following pages – a world where the evidence of past water is incontrovertible, the signs that much of this water remains as ice locked away below the surface convincing, and the possibility that life could once have evolved and might still cling onto existence tantalising.

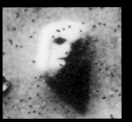
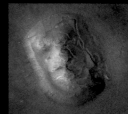
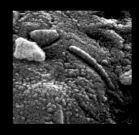

Top: One of the most famous Martian features is the so-called 'Face on Mars', photographed by the Viking mission in 1977, and seen by some people as evidence for ancient Martian intelligence.

High-resolution images taken by Mars Global Surveyor in the 1990s proved what NASA scientists had known from the start – the 'Face' is just a trick of the light and random geology.

In 1997, scientists studying a 'Martian meteorite' – a rock once flung from the surface of Mars to eventually land on Earth – announced tentative evidence for life, including the possible fossils of microscopic bacteria. Most scientists have their doubts, but a manned mission to Mars may be the only way to answer the 'life question' definitively.

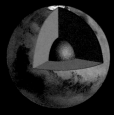

Mars has a simpler structure than the larger terrestrial planets. Its core is relatively small and probably solidified. Above it lies a deep mantle, then the crust, which is thickest beneath the southern highlands.

Distance from sun	Orbital period	Orbital eccentricity	Diameter	Surface gravity	Rotation period	Axial tilt	Natural satellites
227.9	687	0.093	6,780	0.38	24.63	25.19	2
million km	Earth days		km	g	hours	degrees	

227.9

million km
from the Sun

Mars

Marineris hemisphere

Terrestrial planet

This simulated view of the Martian globe, produced by combining images from the Viking orbiter spacecraft, is dominated by the enormous scar of the Valles Marineris, a vast canyon system that stretches almost one third of the way around the planet. On the left limb the huge shield volcanoes that rise in the Tharsis region can be seen.

227.9

million km
from the Sun

Mars

Schiaparelli hemisphere

Terrestrial planet

This view of the Schiaparelli hemisphere shows clear differences in colour between areas of the Martian terrain. The dark surface patches, the only features normally visible from Earth, are thought to form as surface winds move different shades of dust around. Bright areas towards the south indicate carbon dioxide frosts around the Hellas Planitia basin.

227.9

million km
from the Sun

Mars

Cerberus hemisphere

Terrestrial planet

Cerberus hemisphere gets its name from the dark streak just to the left of centre in this image. South of it, the boundary between the rough southern highlands and the much smoother northern plains is clear. More volcanoes, in the Elysium region, can be seen to the north of Cerberus, while the peak of the massive Olympus Mons is partly hidden by cloud at upper right.

227.9

million km
from the Sun

Mars

Syrtis Major hemisphere

Terrestrial planet

The darkest and most prominent region on Mars, Syrtis Major Planum, is a low shield of ancient volcanic rock, streaked with lighter 'shadows' where changes in the landscape affect the prevailing winds. At the southern edge of this view, the Hellas Planitia impact basin is clearly marked by a coating of carbon dioxide frost. All these images use a 3-D projection to simulate the view from orbit.

227.9

million km
from the Sun

Mars

Meridiani Planum

Lowland plain

NASA's mission planners have learned to take advantage of useful side effects. This panorama from the Mars Exploration Rover Opportunity shows the site where its own heat shield hit the surface at the end of its final descent. The impact threw out a useful sample of Martian subsurface soil that would otherwise have been beyond the reach of the rover for analysis.

227.9
million km
from the Sun

Mars

Gusev Crater

Impact crater

This view shows volcanic rock fragments scattered across the interior of Gusev Crater, as seen from near the summit of Husband Hill, one of the Columbia Hills scaled by the Spirit Mars rover. Panoramas such as this combine hundreds of images – in this case the whole scene contains 405 separate pictures captured over six days in late 2005.

227.9

million km
from the Sun

Mars

Columbia Hills

Volcanic hills

The Spirit Mars rover landed in Gusev Crater, a region thought likely to have held a lake in Mars's warm, wet past. However, the soil it found showed few signs of rocks that might have once been submerged. As it rolled through a region called the Columbia Hills, it came across this outcrop of cracked volcanic rock, probably laid down in an eruption from nearby Apollinaris Patera.

million km
from the Sun

Mars

Gusev Crater

Desert

The classic image of Mars is of a dune-covered, red-hued desert similar to this panorama photographed by Spirit in 2006. Ground down by countless millennia of erosion, Martian sand is far finer than typical sand on Earth, and even the thin Martian air can sculpt it into elegantly rippled dune fields. The sands owe their distinctive colouration to their rich iron oxide (rust) content.

227.9

million km
from the Sun

Mars

Endurance Crater

Impact crater

The Opportunity rover struck geological gold when it landed near the 30-metre (100-foot) Endurance Crater. As it peered over the edge, it discovered layers of sedimentary rock around the crater rim – rock that could only have formed underwater. The spectrometer it used for chemical analysis also discovered pebbles of the mineral haematite, which also normally forms underwater.

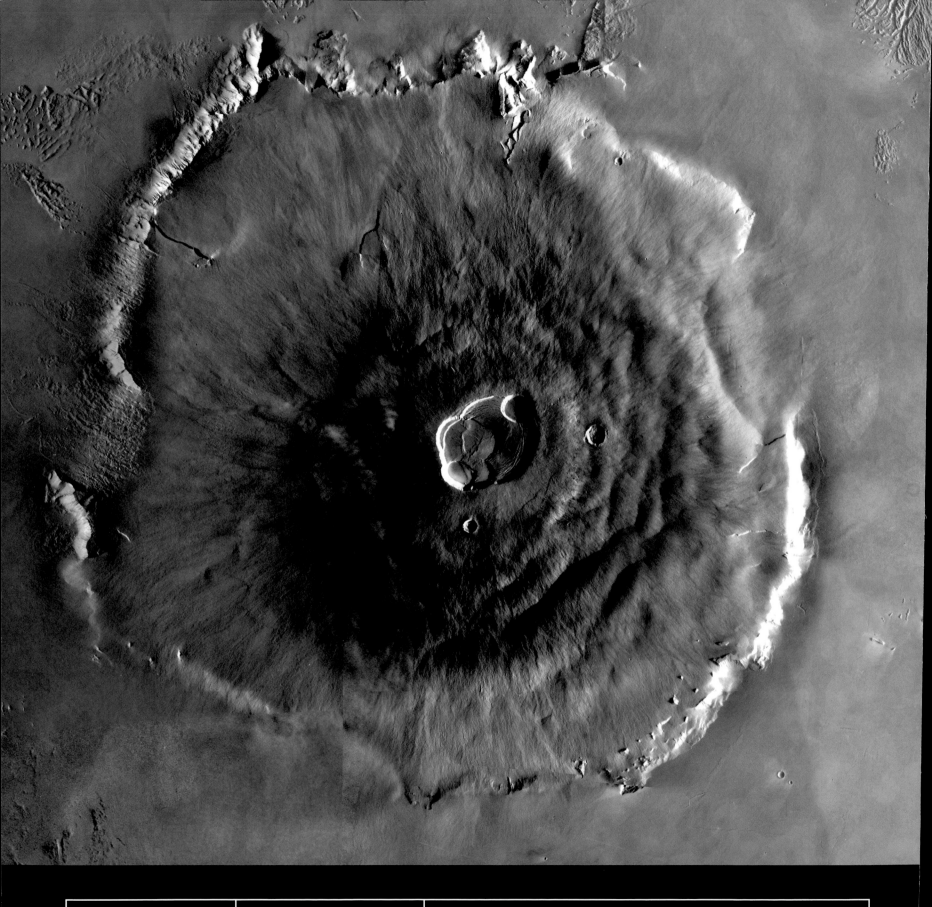

million km
from the Sun

Mars

Olympus Mons

Volcanic mountain

Towering volcanoes are without doubt the most famous Martian features, of which Olympus Mons is by far the largest. This enormous volcanic shield, shown here in a Viking orbiter photograph, rises to 27 km (17 miles) above the planet's mean surface level – almost three times the height of Mount Everest. This makes Olympus Mons the largest volcano in the entire Solar System.

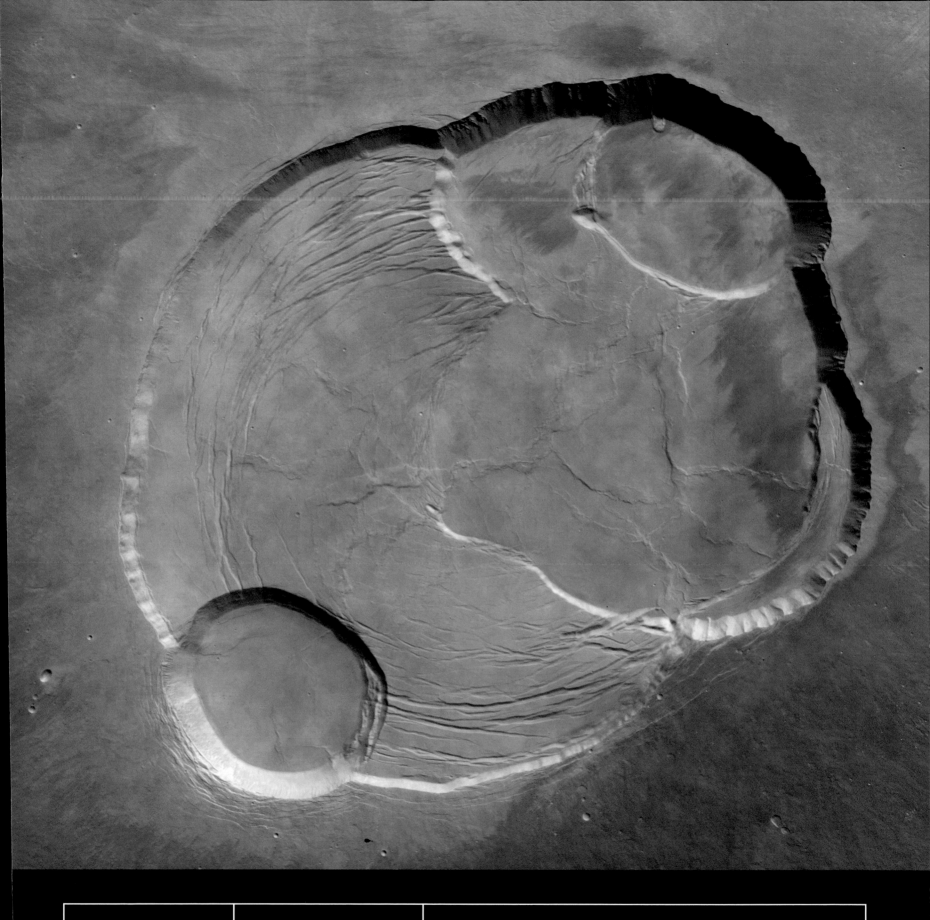

227.9

million km
from the Sun

Mars

Olympus Mons caldera

Volcanic mountain

The complex caldera at the peak of Olympus Mons was formed by several eruptions at different times. The large central region, marked by concentric cracks around its edges, is the youngest. The walls surrounding the caldera plunge some 6 km (3.7 miles) to its subsided floor. Astronomers believe that the volcano's eruptions finally spluttered to a halt around 30 million years ago.

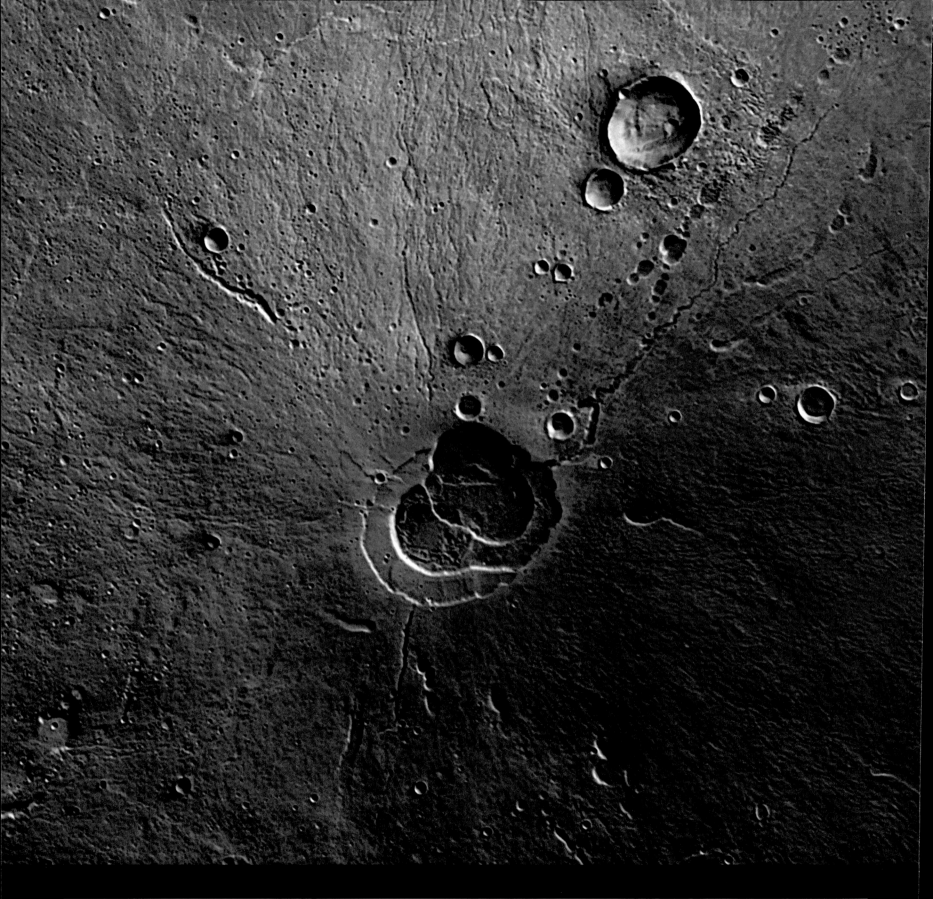

227.9

million km
from the Sun

Mars

Hecates Tholus

Volcanic mountain

Hecates Tholus is a large volcano in the Martian tropics – as with Olympus Mons, the multiple calderas at its summit probably formed during different periods of activity. Most of the lava flows here are very ancient and have been heavily cratered since they formed, but some areas are almost pristine and probably young. Some Martian volcanoes might still be active today.

227.9

million km
from the Sun

Mars

Albor Tholus

Volcanic mountain

Close to Hecates Tholus in the Elysium region of Mars lies Albor Tholus, a mid-sized Martian volcano. This exaggerated 3-D view from Mars Express focuses on the caldera, 30 km (19 miles) across and 3 km (1.9 miles) deep. The volcano as a whole is some 160 km (100 miles) wide. On the left of the caldera, Martian dust seems to be sifting down to the crater floor.

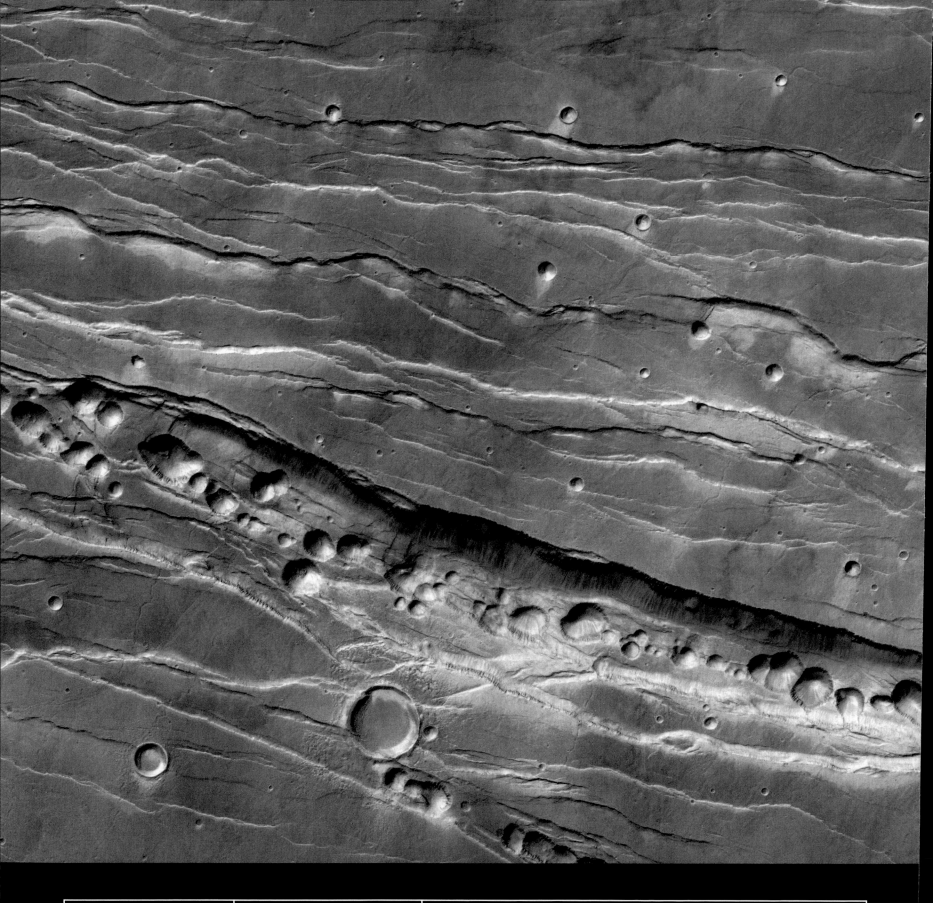

227.9

million km
from the Sun

Mars

Phlegethon Catena

Crustal fault

The long crater chain of Phlegethon Catena might at first appear to be the result of a string of impacts, perhaps from a broken-up comet, but appearances are deceptive. Its location in the midst of 'graben' trenches at the edge of the Alba Patera volcano suggests that the circular pits probably formed through subsidence, perhaps as the volcano melted away pockets of underground ice.

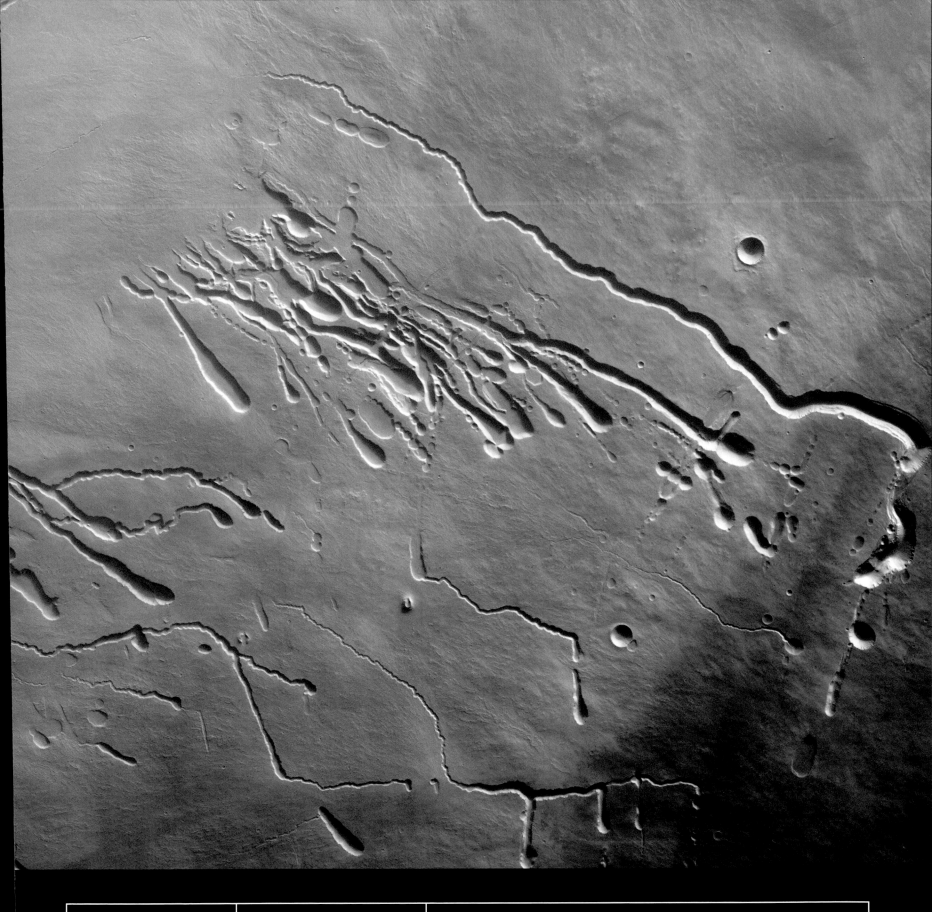

million km
from the Sun

Mars

Pavonis Mons rivulets

Volcanic lava tubes

At first glance these rivulets in the Martian landscape look like evidence of water, but once again they are really the result of volcanism at work. These are collapsed lava tubes on the flank of Pavonis Mons – the traces of channels where lava once flowed through underground tunnels beneath a thin solidified crust. As the tubes emptied, the surface above them collapsed to form valleys.

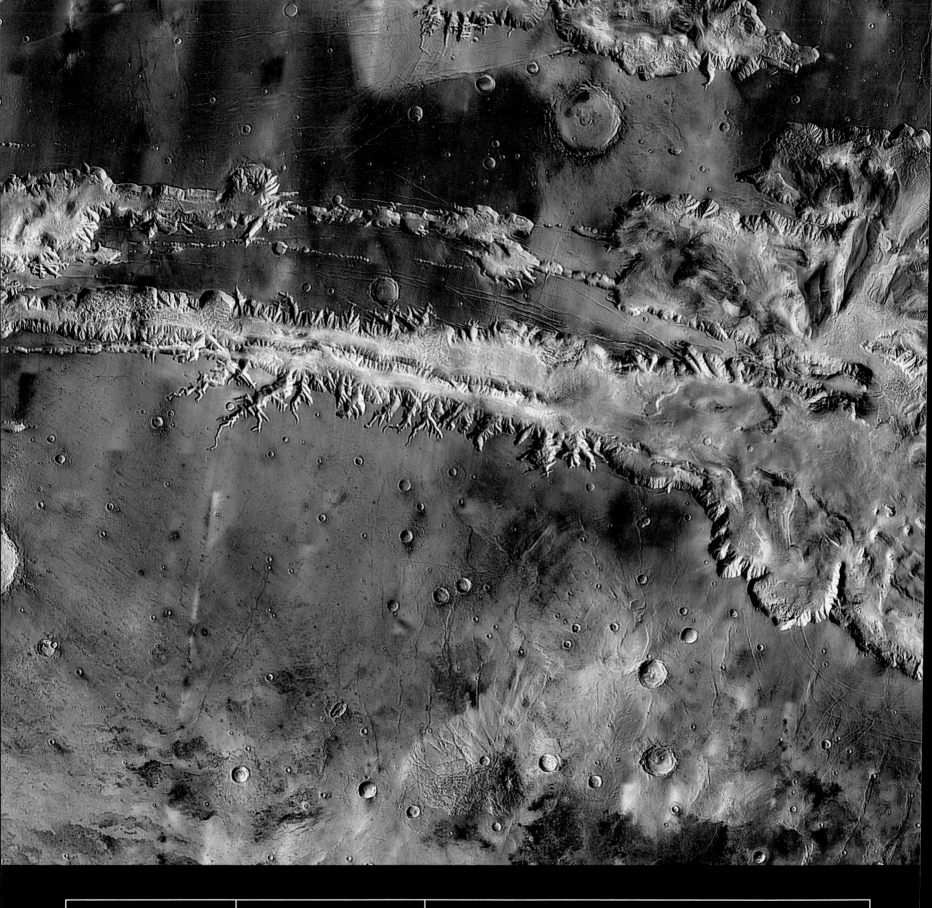

Mars

227.9

Valles Marineris

Crustal fault

The Valles Marineris cuts 9 km (5.5-miles) into the Martian crust south of the volcanic Tharsis region. Its scale is hard to comprehend – at 4,000 km (2,500 miles) long it runs almost one fifth of the way around the planet, with multiple canyon systems giving it a maximum width of 700 km (435 miles). In comparison, Earth's Grand Canyon is only one tenth the length and one fifth the depth.

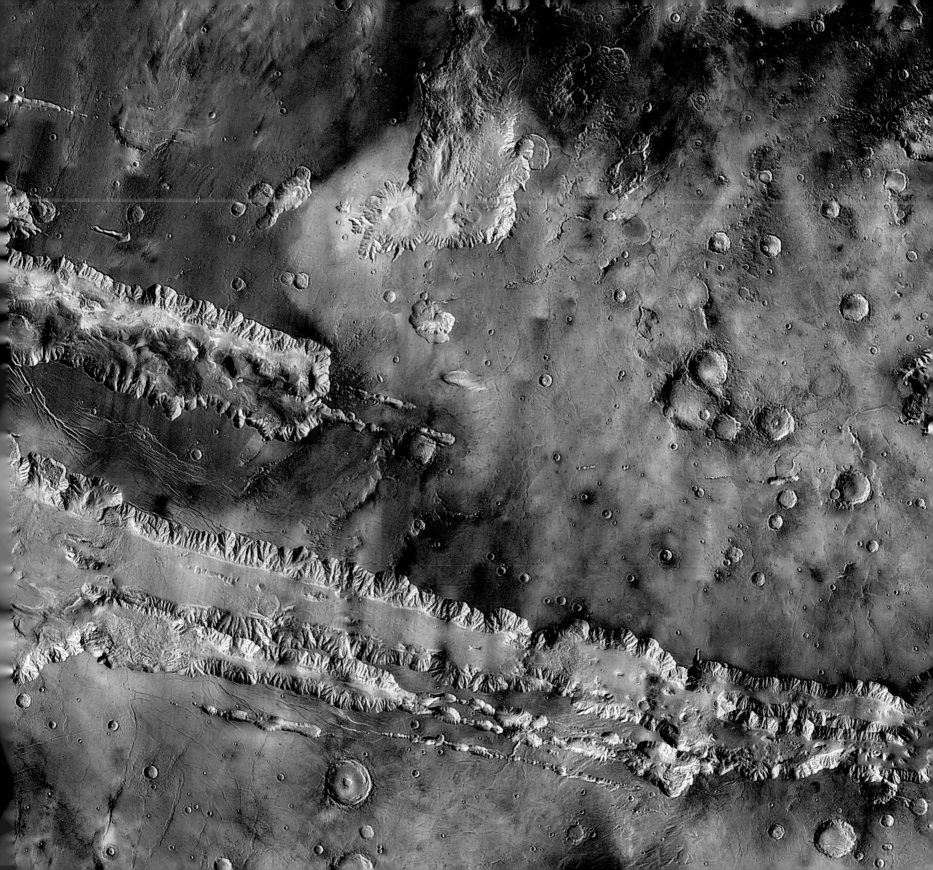

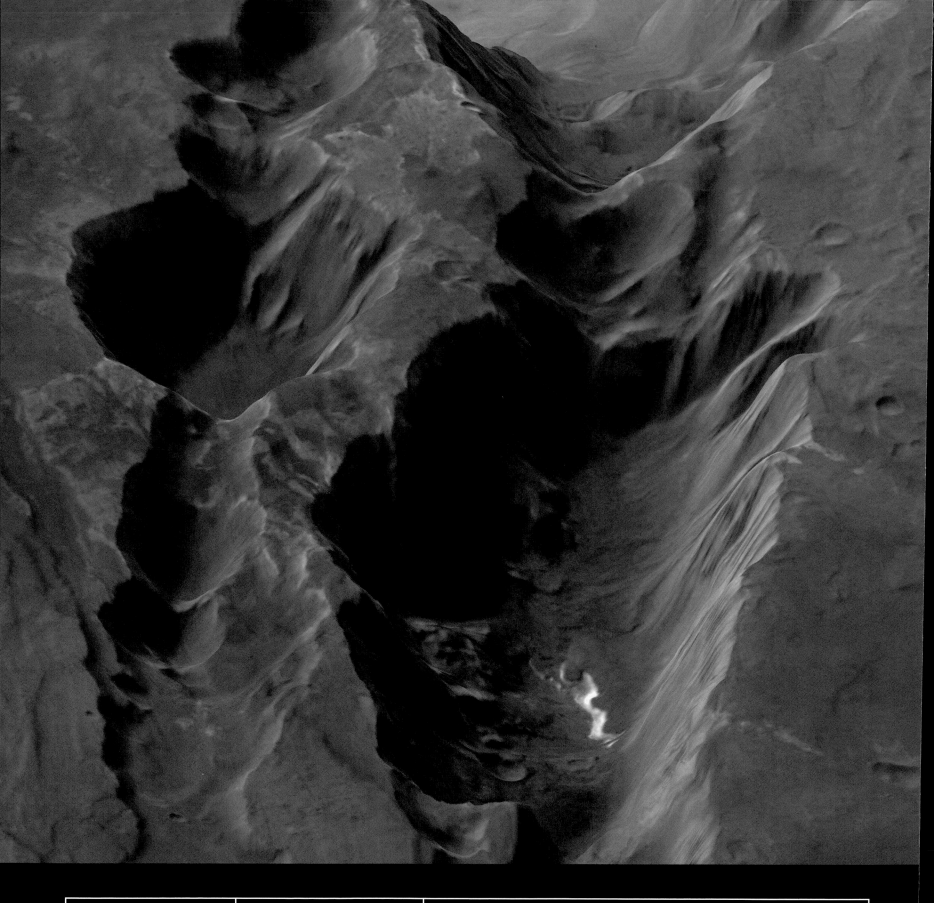

227.9

million km
from the Sun

Mars

Coprates Chasma

Crustal fault

The main trench at the eastern end of the Valles Marineris, Coprates Chasma (on the right in this Mars Express image) is a huge trough ranging from 60-100 km (36-62 miles) across and up to 9 km (5.5 miles) deep. To the left lies Coprates Catena, a second trench apparently formed as the crust collapsed to form a series of pits, perhaps undermined by volcanic activity or water erosion.

227.9

million km
from the Sun

Mars

Ophir Chasma

Crustal fault

The central regions of the Valles Marineris, around Ophir Chasma, are a chaos of broad trenches separated by ridge-like high plateaus. There are clear signs that narrower channels have collapsed into the valley below. These collapses might have been triggered by catastrophic mudslides, perhaps as freshly melted water, warmed by the Tharsis volcanoes, burst from the valley walls.

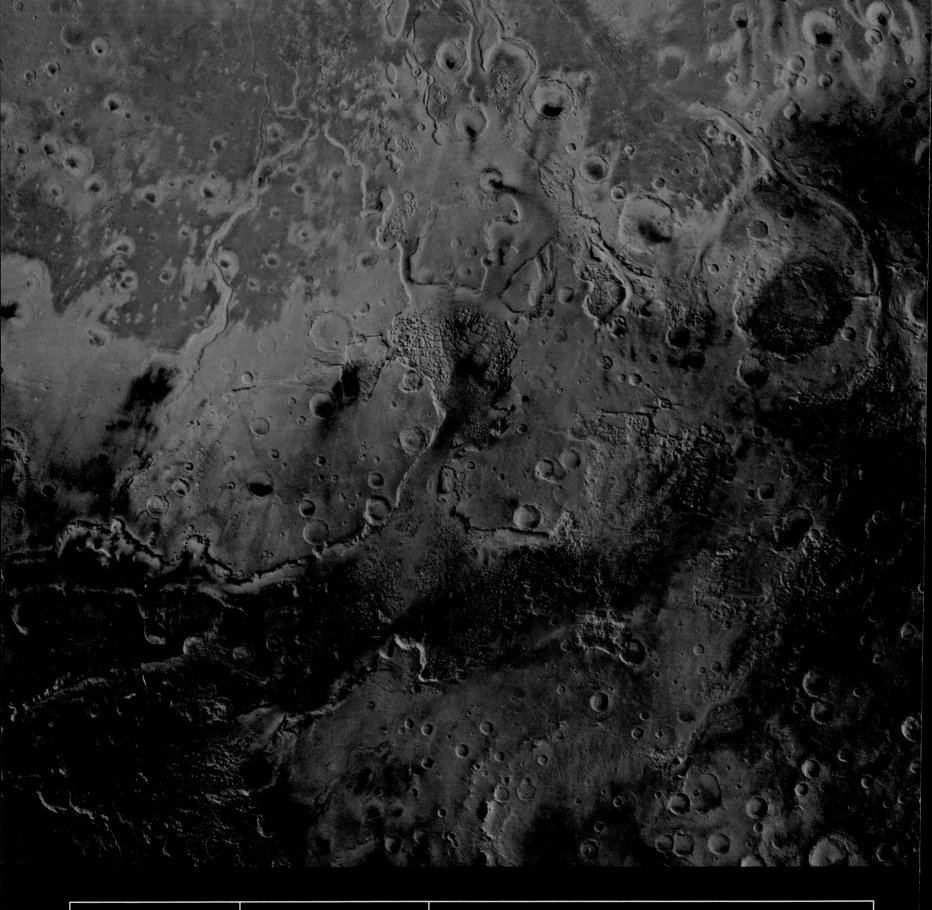

227.9

million km
from the Sun

Mars

Chryse Planitia

Erosion feature

The scars of catastrophic flooding are seen at work in the Chryse outflow region to the northeast of the Valles Marineris. The teardrop-shaped islets of Chryse show where torrential floods swept through the landscape. On Earth, similar scars mark the regions where flood waters burst from behind glacial dams at the end of the last ice age.

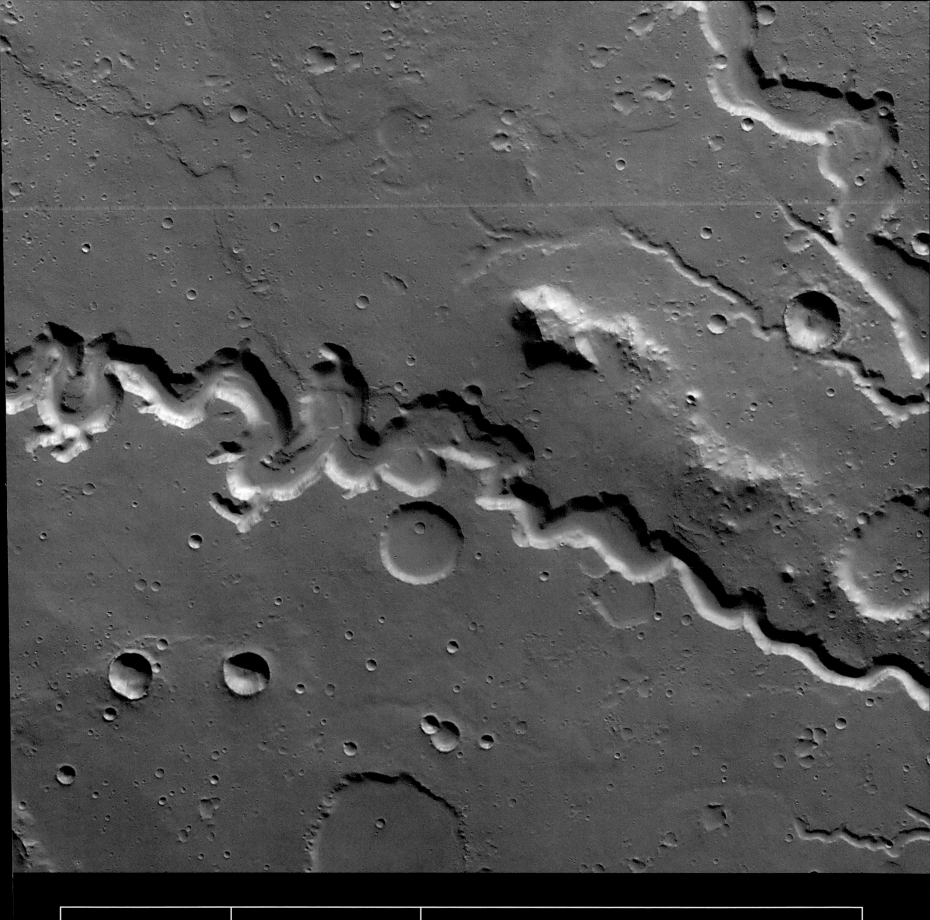

227.9

million km
from the Sun

Mars

Nanedi Vallis

Erosion feature

Nanedi Vallis and other channels like it are evidence of water at work over much longer periods of time – they resemble winding valleys formed on Earth by the flow of water over many thousands of years, if not longer. Although features like this might have formed by some other means, the conclusive evidence of water found elsewhere suggests that rivers are the right explanation.

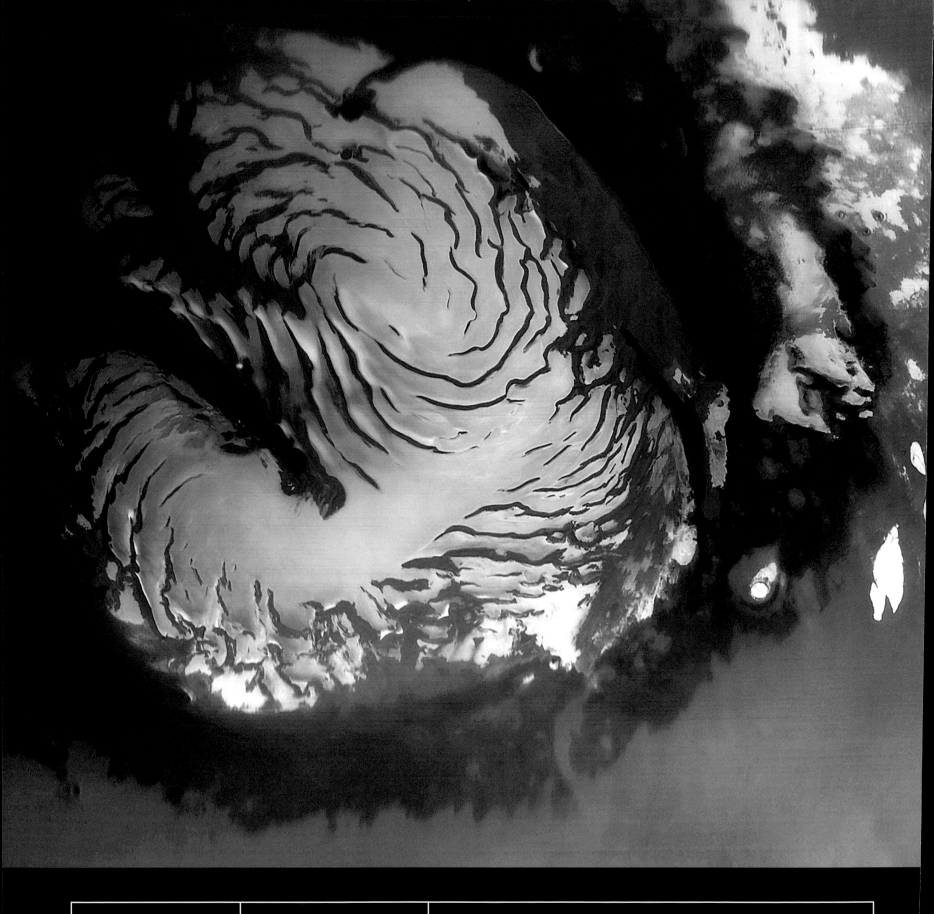

million km
from the Sun

Mars

North pole

Polar cap

The Martian polar caps rise to heights of several kilometres above their surroundings, and have surprisingly complex, swirling shapes. This is the larger and more impressive north pole, photographed by Mars Global Surveyor. Unlike the south pole, its upper layers of carbon dioxide frost, which grow and shrink with the Martian seasons, sit on top of a more permanent cap of water ice.

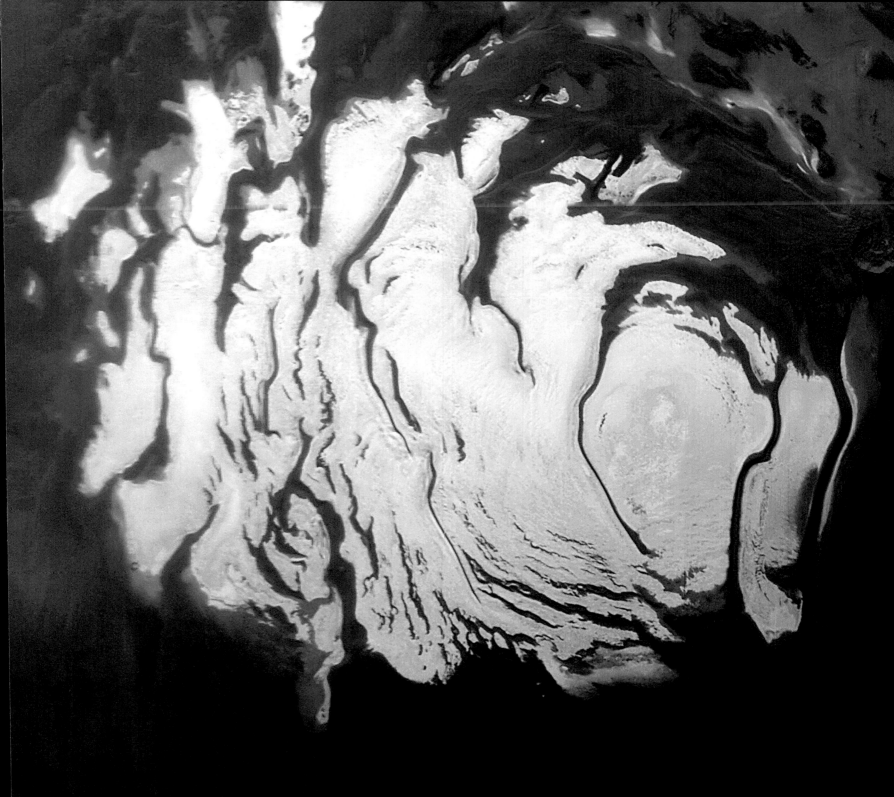

227.9

million km
from the Sun

Mars

South pole

Polar cap

The south pole is dominated by carbon dioxide ('dry ice'), its swirls and ice-free
gullies sculpted by prevailing winds that control where ice settles each winter.
In spring, when the pole is exposed to sunlight again, the ice sublimes, stirring
winds as fast as 400 kph (250 mph). Overall, the southern ice cap has receded
in recent years – dramatic evidence of a changing Martian climate.

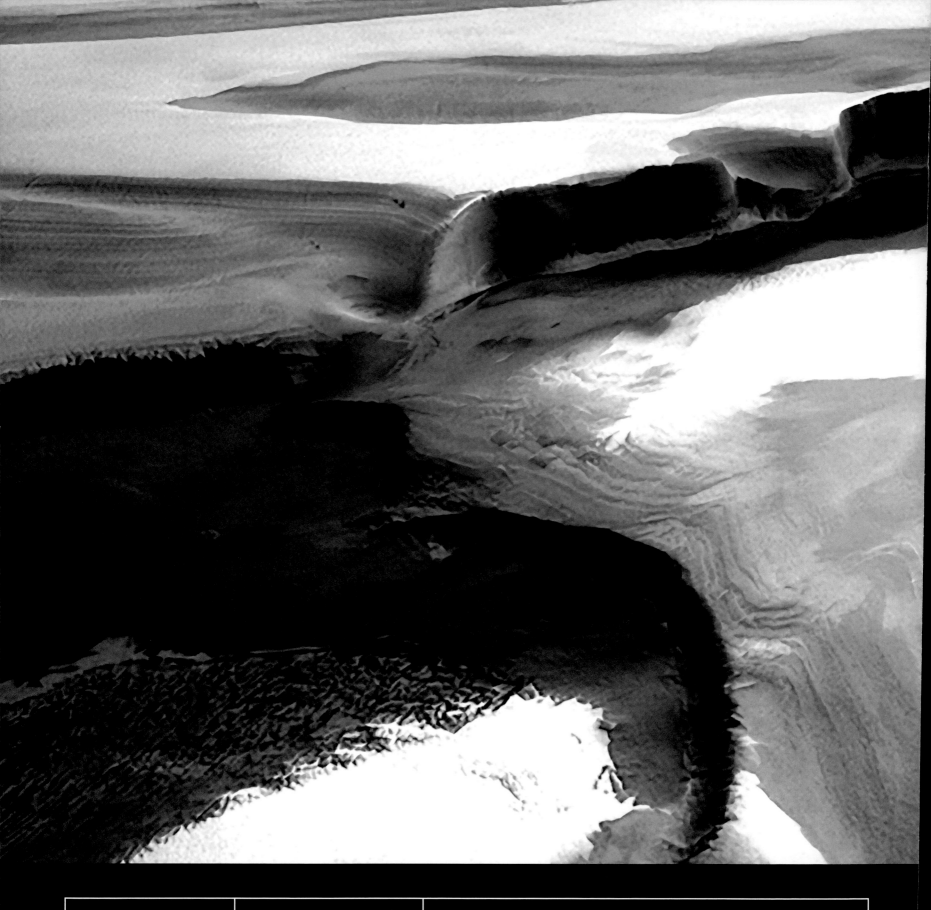

227.9

million km
from the Sun

Mars

Polar layers

Ice feature

Both poles display impressive terraced terrain – the cliffs in this image from Mars Express are up to 2 km (1.2 miles) high. Each new winter's frost deposits carry with them fine dust from the Martian atmosphere, which remains when the ice itself evaporates in spring. Over the aeons, these thin layers of dust have accumulated to form the terraces.

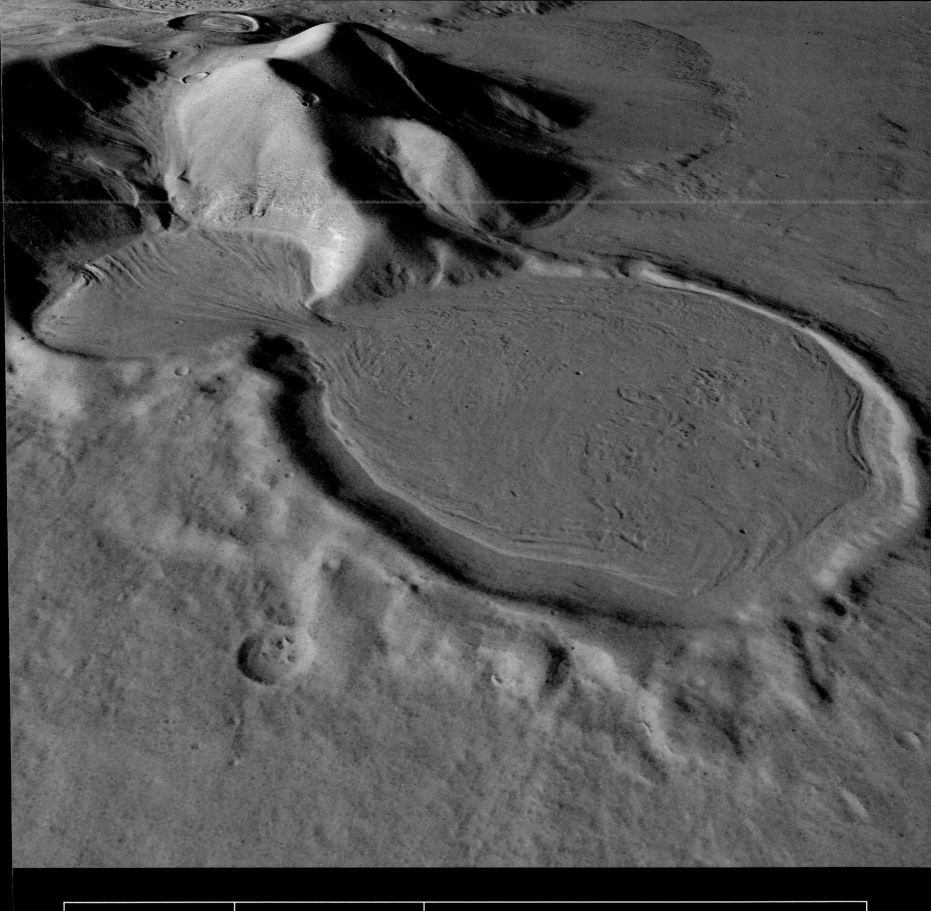

Mars

Hourglass Crater

This pair of neighbouring craters lies in the southern hemisphere's Hellas Planitia (see p.122). High-resolution 3-D images from Mars Express show where a stream of debris has flowed from the smaller, higher crater into its

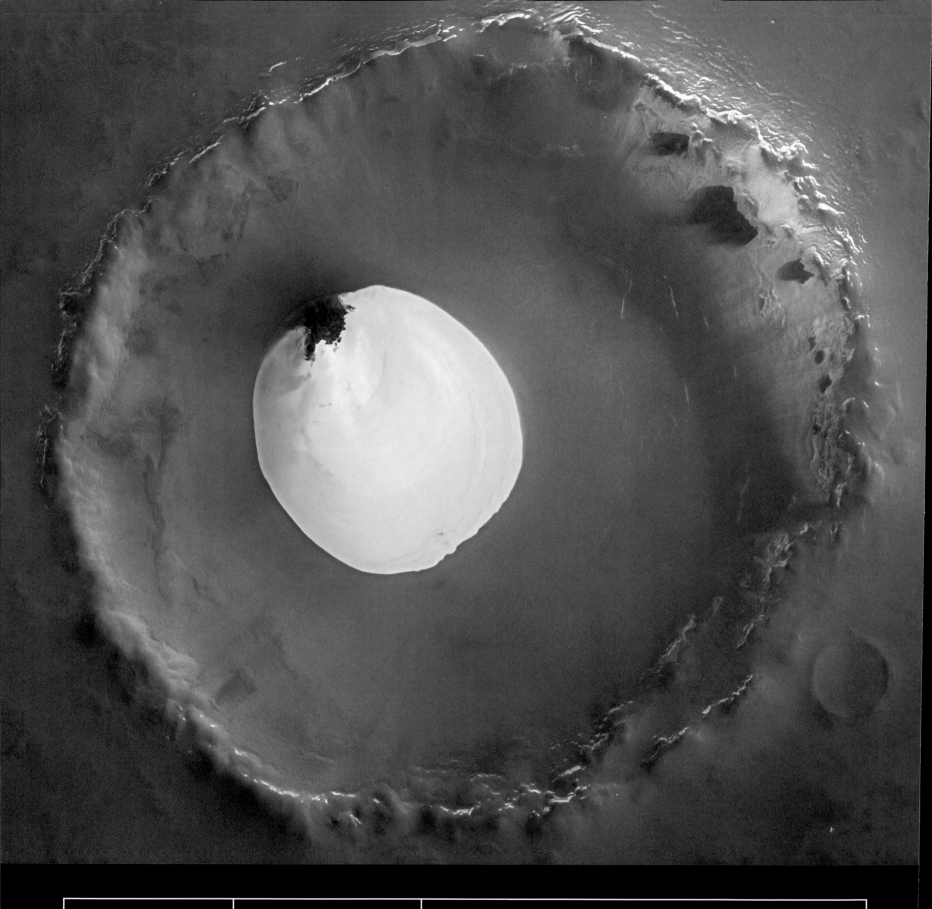

227.9

Mars

Ice crater

Impact crater

A few isolated patches of ice survive at some distance from the main polar caps, and the most spectacular of these is undoubtedly the ice-filled crater discovered by the European Mars Express spaceprobe in 2005. Water ice probably survives here because the crater's walls help to shield it from the Sun for much of the Martian day.

Mars

Pack ice

Ice feature

Much of the terrain surrounding the polar caps resembles the permafrosts of Earth – tundra-like regions where the ground remains frozen all year round. On longer timescales, cycles of thawing and refreezing have caused the ground to split into blocks, creating geometric patterns such as these. It's thought that Mars is currently in the cold phase of a multi-million year climate cycle.

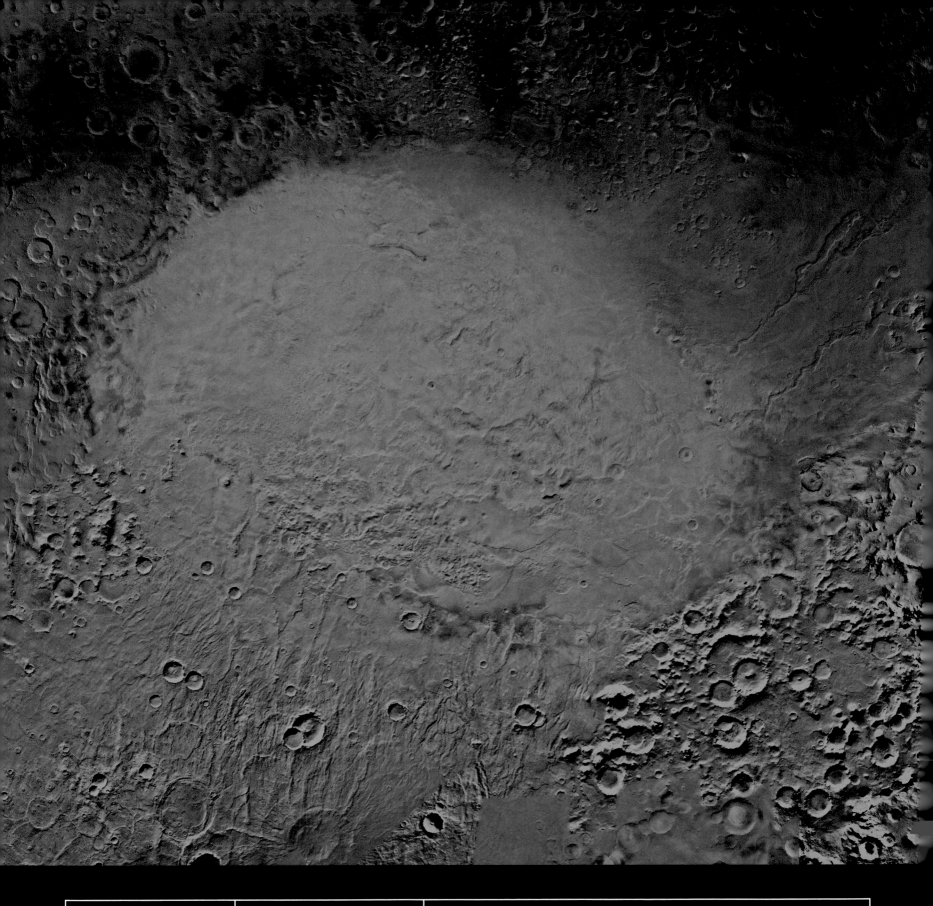

Mars

Hellas Planitia

Impact basin

Hellas is the largest impact basin on Mars, a great sandy desert in the midst of the southern highland regions, 2,200 km (1,365 miles) across. In the 4 billion years since it formed, the basin has been modified by the action of lava, water and wind, but parts of the crater rim still survive, and concentric cliffs in the surrounding highlands trace out what must once have been a huge ring system.

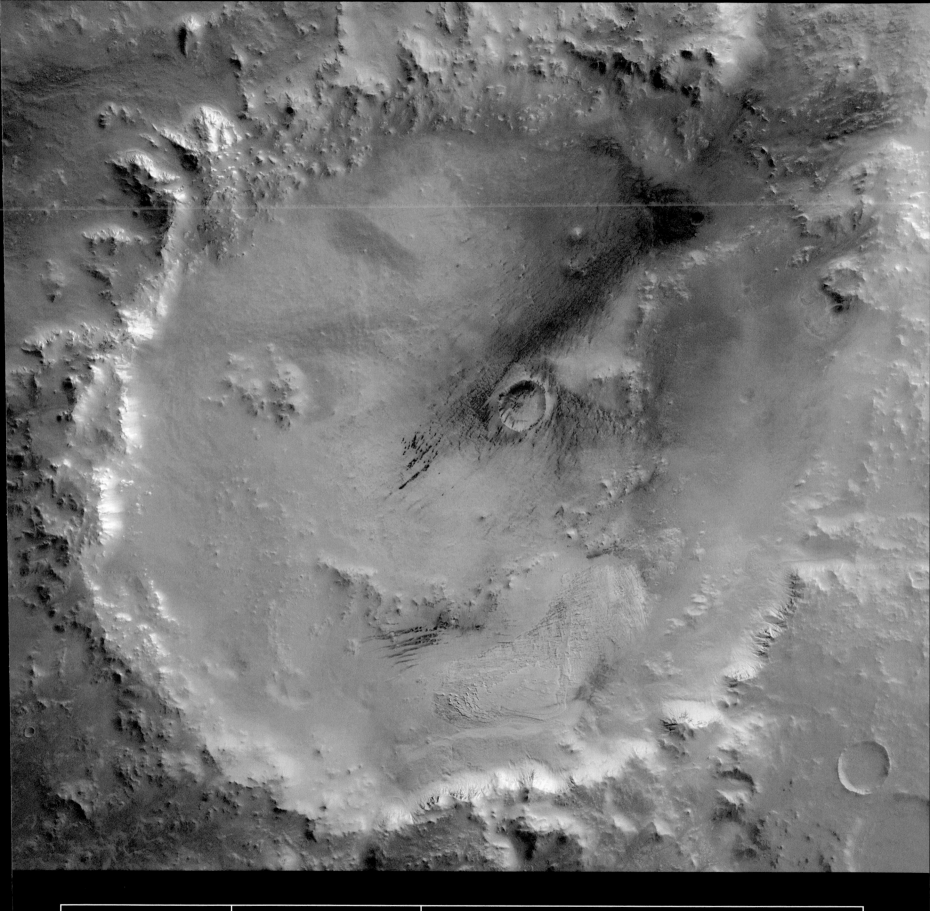

million km
from the Sun

Mars

Galle Crater

Impact crater

While the more famous 'Face on Mars' is now established as nothing more than a trick of the light playing on a rocky mesa (see p.88), the 230-km (143-mile) Galle crater provides a cartoonish, but less fleeting, 'happy face' replacement. The crater floor is covered in dune fields, and shows the dark tracks of Martian dust devils (see p.127)

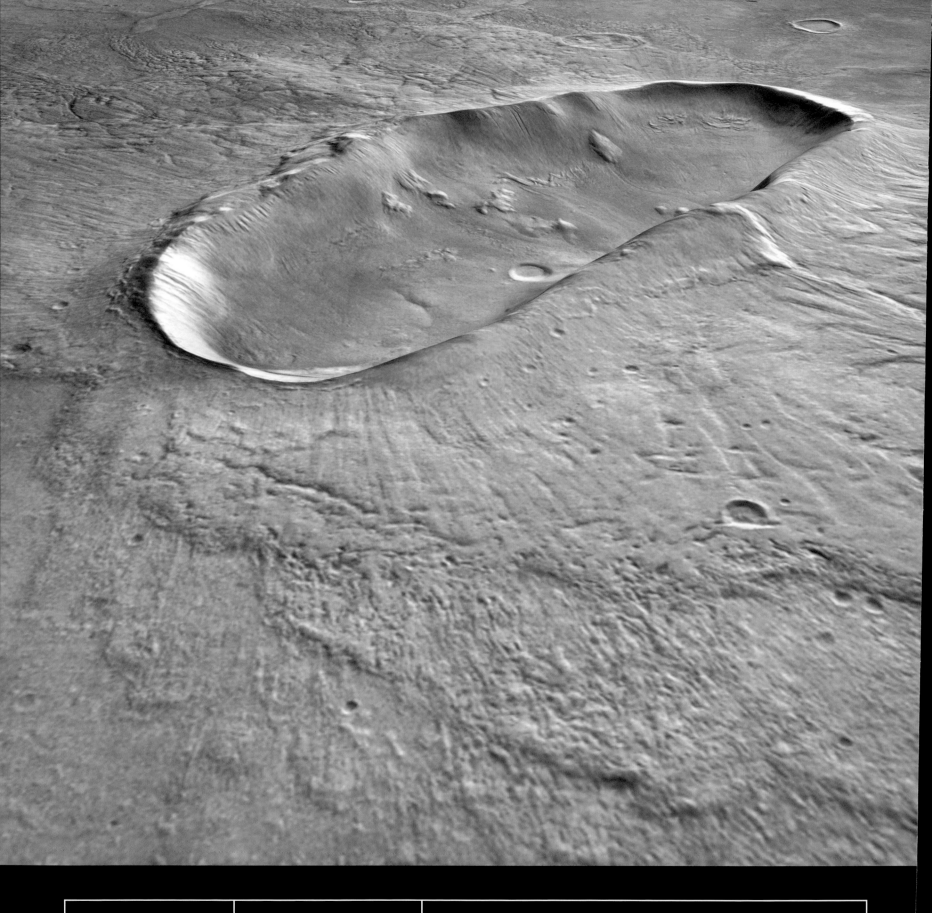

227.9
million km
from the Sun

Mars

'Butterfly' crater

Impact crater

This so-called 'butterfly' crater 24.4 km (15.2 miles) long and 11.2 km (7 miles) wide, is a rare elliptical crater formed when a meteorite hit the Martian surface at a very low angle. Two huge lobes of ejecta material can still be seen to its northwest and southeast, but the crater and its surroundings seem to have been partially re-covered by lava flows from nearby volcanoes.

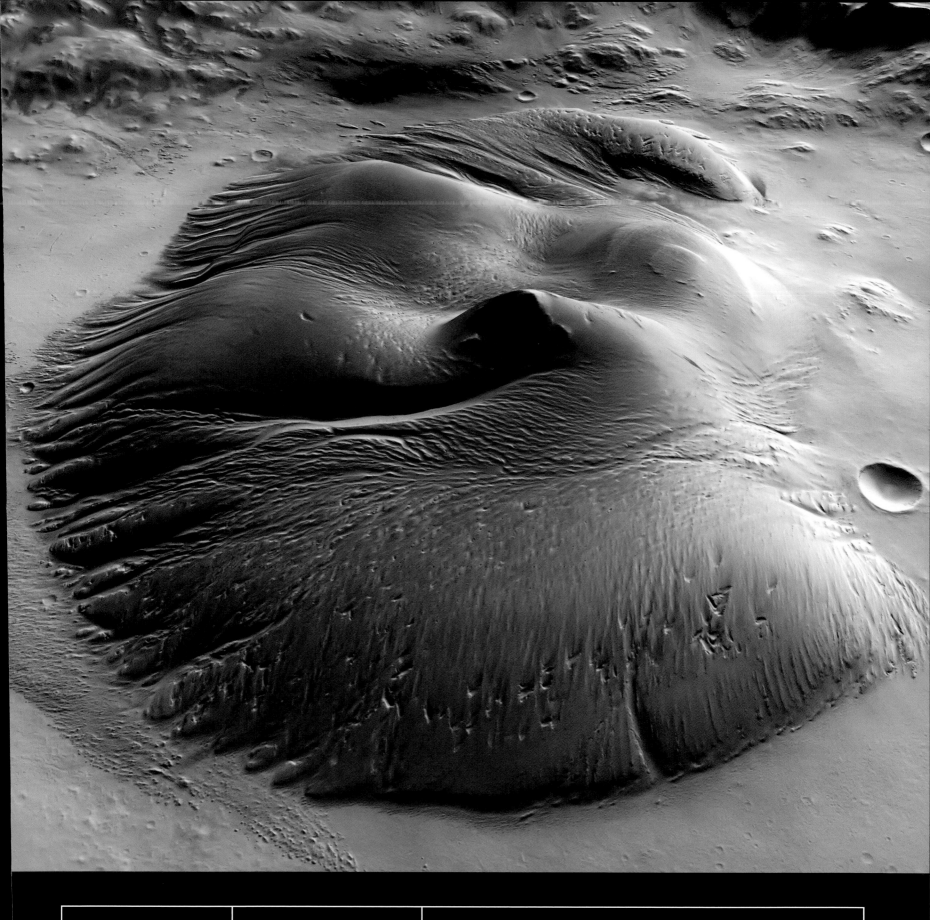

227.9

million km
from the Sun

Mars

Nicholson crater

Impact crater

The floor of the 100-km (62-mile) Nicholson Crater is dominated by a raised feature, 55 km (34 miles) long. The high point at the centre is clearly the central peak common in large craters, formed as the crater floor 'rebounds' during its formation, but the origin of the rest of the material is still puzzling. One thing is clear, though – it has been heavily eroded since it formed.

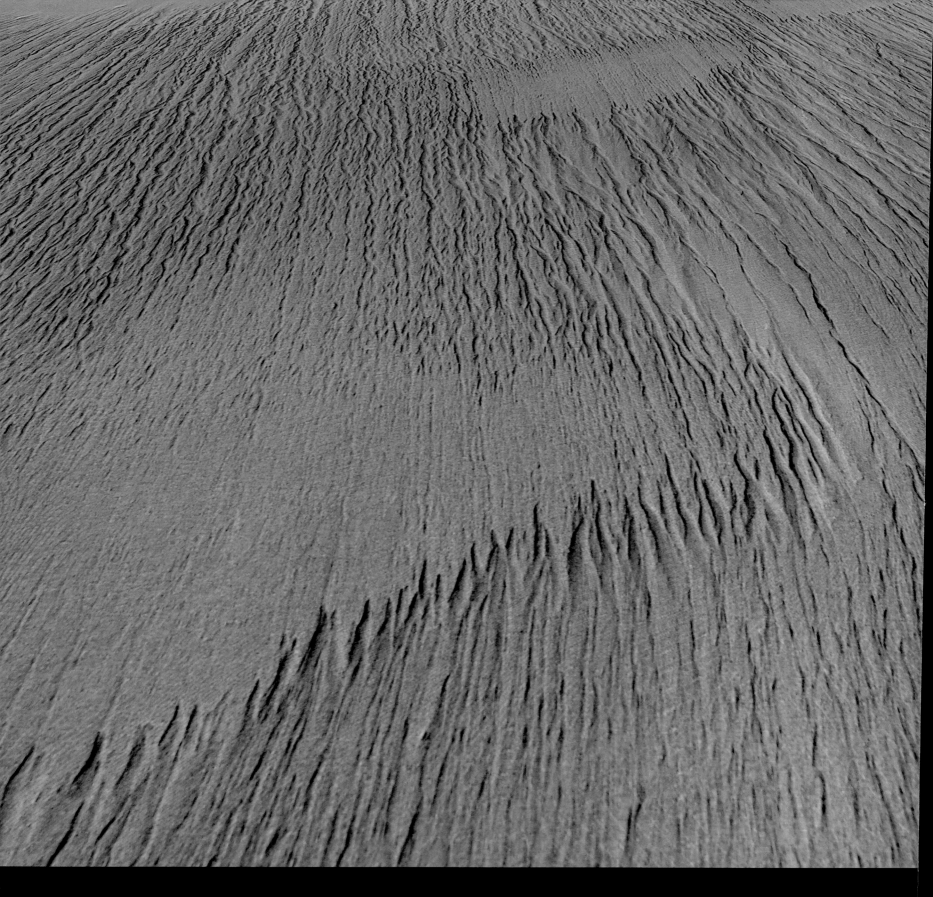

227.9

million km
from the Sun

Mars

Martian yardangs

Erosion feature

These parallel ridges south of Olympus Mons are the Red Planet's equivalent of Earth's yardangs – features created by the effect of continual wind blowing in one direction across a desert. Here the fine Martian sands have gradually eroded the soft bedrock to create trenches that run for dozens of kilometres, and are only interrupted by the occasional outcrop of more resistant material.

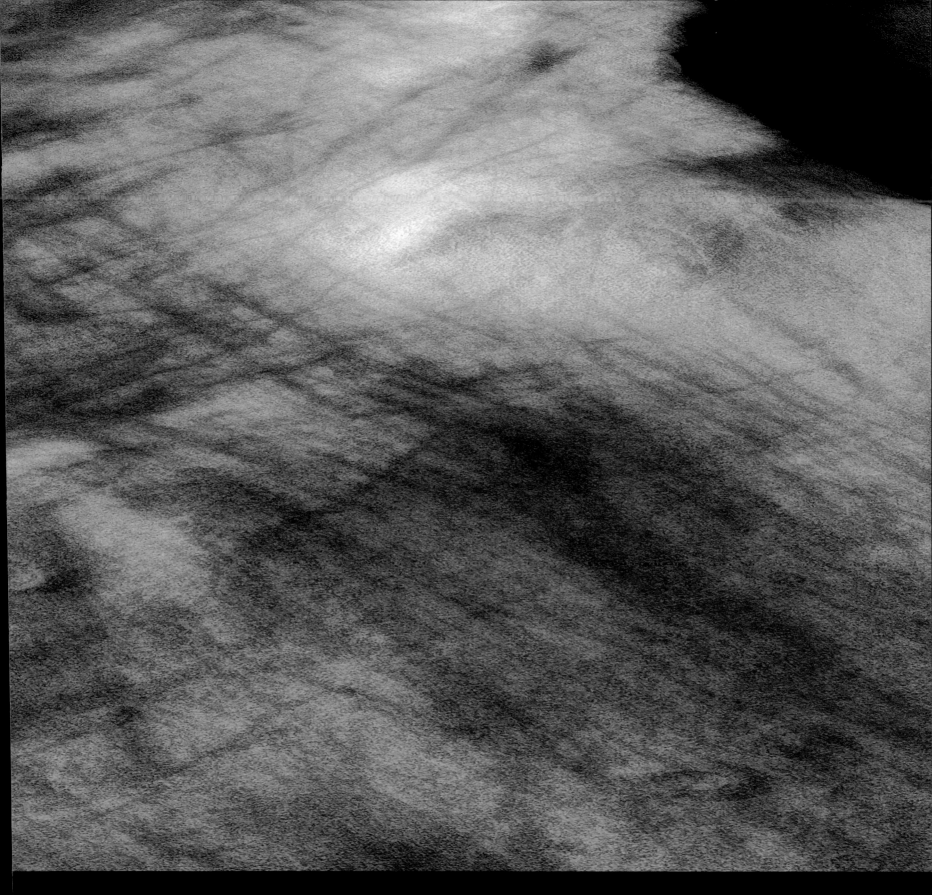

million km
from the Sun

Mars

Dust devil tracks

Erosion features

Though rarefied, the Martian atmosphere is still capable of whipping up the fine, light Martian sands to form dust devils. These mini-tornadoes have been photographed from orbit and on the surface. As they clear away fresh dust in their path, they reveal darker soil beneath, creating random 'scribbles' across the plains of Mars or, as here, more orderly criss-cross patterns.

227.9

million km
from the Sun

Mars

2001 dust storms – before

Weather feature

Mars is famous for the ferocious dust storms that periodically sweep across its surface, sometimes obscuring the entire planet from view and lasting for several weeks. Although the atmosphere is much thinner than Earth's and so the high winds have less force, they can still have an impressive effect because the dust-like Martian sand is so easily stirred up and so slow to settle.

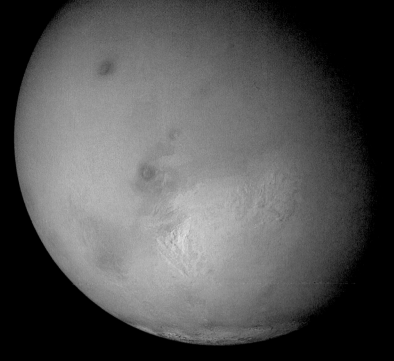

227.9

million km
from the Sun

Mars

2001 dust storms – during

Weather feature

The pair of Mars Global Surveyor images shown here and opposite chronicle the spread of a global dust storm around Martian perihelion in 2001. Mars has a relatively eccentric orbit, and as its distance from the Sun varies so does the heat available to power its weather systems. As a result, though local dust storms are relatively common, global ones only occur during Martian perihelion.

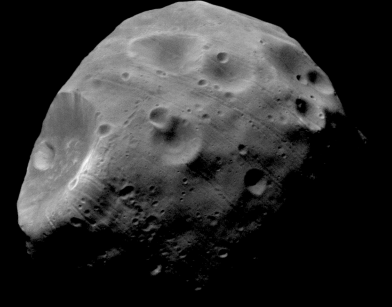

9.4

thousand km
from Mars

Phobos

Length: 26.8 km

Rocky captured satellite

Mars has two small moons, named after the ancient Roman god's dogs of war. Phobos is the larger and closer to Mars. Its surface is heavily cratered and also displays strange parallel 'scrapes'. As Phobos circles the planet in just 7 hours 39 minutes, it clips the upper atmosphere. In around 40 million years, its orbit will decay to instability and it will crash into Mars itself.

23.4

thousand km
from Mars

Deimos

Length: 15 km

Rocky captured satellite

Deimos is the smaller and more distant Martian moon, orbiting its parent planet in 30 hours 18 minutes. Both Phobos and Deimos are thought to be captured asteroids, though the reddish colouring of their surfaces suggests that they might be stray 'Trojans' from around Jupiter's orbit (see p.132) rather than refugees from the main belt.

Among the Asteroids

Between the orbits of Mars and Jupiter lies the domain of the asteroids, a realm of numberless small, rocky worlds, the largest still far smaller than our own Moon. These rocks were low on the list of priorities for early space exploration, and at first were visited as mere afterthoughts for more ambitious missions. Yet the more we discover about the asteroids, the more mysteries they throw up. In addition, we now realise that some asteroids present a serious threat to Earth, and so the quest to learn more about them has taken on an added urgency.

The vast majority of asteroids orbit in the 'Main Belt', extending from 28 million km (17.5 million miles) beyond the orbit of Mars, out to around 598 milliom km (372 million miles), 180 million km (112.5 million miles) inside the orbit of Jupiter. Their distribution offers an important clue to the forces at work in this region of space – the belt's inner edge closely follows the somewhat elliptical orbit of Mars, while the outer edge steers well clear of Jupiter. Within the belt there are 'Kirkwood gaps' where no asteroid orbits are found – these correspond to regions where an asteroid's orbital period would regularly bring it into close alignment with Jupiter. Any asteroid straying into these gaps would suffer repeated gravitational 'kicks' that would eventually force it into a different orbit.

Despite the fact that the main belt contains an estimated 200 million asteroids more than 100 metres (300 ft) across, it fills such a huge region of space that individual asteroids are well scattered within it. The first spaceprobes to the outer planets cut across the belt without ever encountering an asteroid, and the Jupiter-bound Galileo probe had to be specifically directed to follow a flight path incorporating close encounters with Ida and Mathilde. Even these first two encounters revealed surprising variation among the asteroids – Ida proved to be an ovoid world with its own small satellite, Dactyl. Based on the way its gravity affected Galileo's flightpath, it is made of

solid rock, though with lower density than the highly compressed rocks on Earth. Mathilde meanwhile, was a larger world, very roughly circular, but with far less mass, suggesting that it is little more than a 'rubble pile' – a cluster of rocks loosely held together by gravity, but with much of its interior nothing more than empty space. Most asteroids conform to one of these two main types, but there are notable exceptions, probably the strangest of which is Vesta, the third largest asteroid (see p.137).

Not all asteroids are confined to the main belt, however. On the sunward side there are numerous Near Earth Asteroids (NEAs), refugees most likely ejected from the Kirkwood Gaps into orbits that bring them within that of Mars, and sometimes into orbits that cross Earth's own. Depending on their orbit, they are sometimes classified as Amors (orbiting beyond Earth, in Mars-crossing orbits), Apollos (with Earth-crossing orbits, but spending most of their time beyond Earth), and Atens (with orbits that are largely inside that of Earth). On the far side of the belt, Jupiter shares its orbit with two clusters of asteroids known as the 'Trojans', whose orbits keep them either far ahead or far behind the giant planet itself, and safe from its gravitational influence.

A NEA called Eros has the distinction of being the most studied asteroid. The Near Earth Asteroid Rendezvous probe NEAR-Shoemaker entered orbit around this Earth-crosser on the appropriate date of February 14, 2000, surveying it for a whole year before finally being guided to a gentle touchdown on its surface. It found a roughly potato-shaped rock some 31km (19 miles) long, with surprisingly strong gravity that indicates Eros has a mostly solid interior. On a celestial timescale, NEAs such as Eros are thought to have relatively short lifespans – once ejected from the main belt, they will inevitably have gravitational encounters with the inner planets, which may result in a direct collision, or send them spiralling further in towards the Sun.

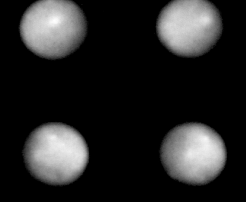

Ceres, the largest asteroid of all, orbits the Sun in 4.6 years, around the middle of the main belt. We know little about it, but these blurred Hubble Space Telescope photographs prove that it has a generally dark surface, and a mysterious bright spot close to its equator. It is also large enough, at 960 km (596 miles) across, to pull itself into a spherical shape.

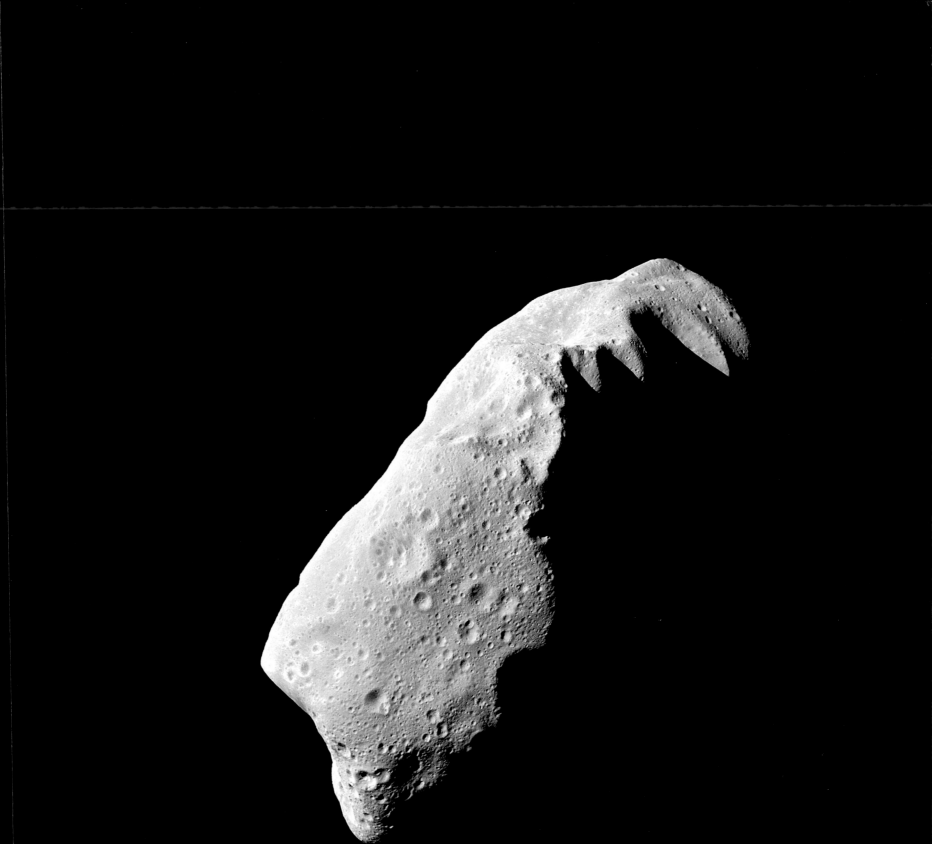

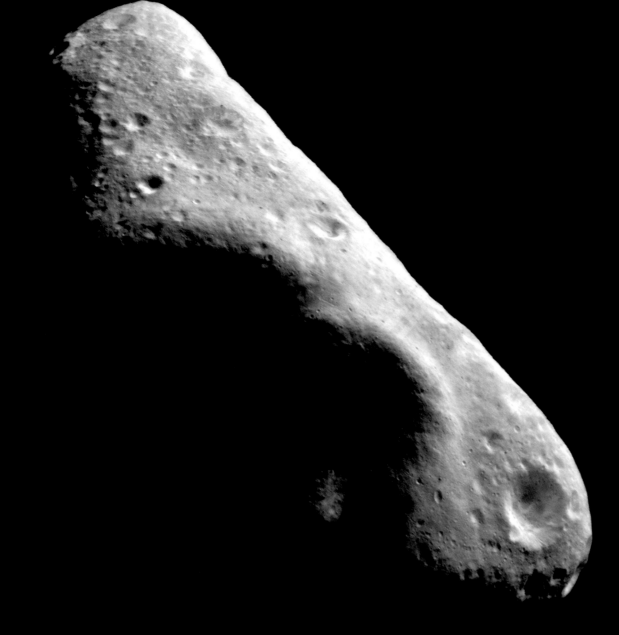

million km
from the Sun

433 Eros

Length: 31 km

Near Earth Asteroid

The most intensively studied asteroid, Eros was orbited for an entire year by the Near-Earth Asteroid Rendezvous probe NEAR-Shoemaker. Eros is one of many Near-Earth Asteroids that orbit inside the main belt – it circles the Sun once every 1.76 years, and there is a one in ten chance that its orbit will intercept Earth's in the next million years.

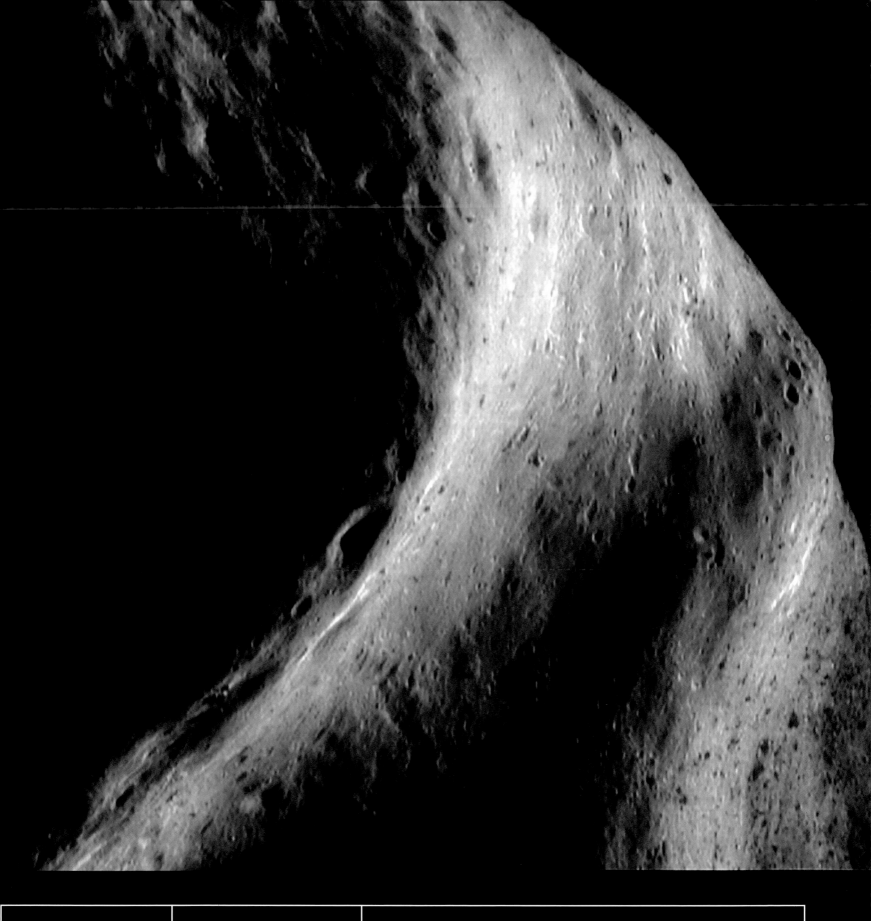

433 Eros

'The Saddle'

Impact crater

Eros' most distinctive feature is the saddle-shaped hollow called Himeros. This depression in Eros's convex side is almost certainly an ancient impact crater. Boulders up to 50 m (165 ft) across can be seen scattered to its right. One of the biggest surprises about Eros was how much erosion had softened the landscape, presumably a result of endless bombardment by micrometeorites,

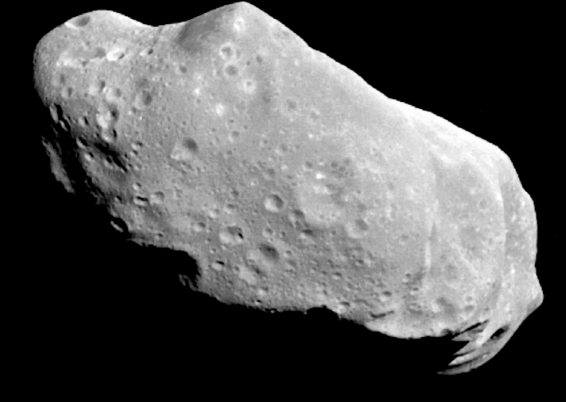

428
million km
from the Sun

243 Ida

Length: 60 km

Main belt asteroid

Our first really good look at an asteroid came when the Jupiter-bound Galileo probe sped through the asteroid belt in 1993. Galileo's photographs of 243 Ida revealed an irregularly shaped lump of rock, some 54 km (33.5 miles) long, and composed mostly of carbon-rich minerals. Surprisingly, Ida turned out to be orbited by a small moon, 1.4-km (0.9-mile) Dactyl.

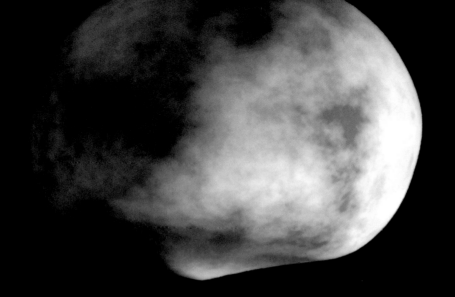

353
million km
from the Sun

4 Vesta

Diameter: 560 km

Main belt asteroid

Asteroid 4, Vesta is the third largest of all asteroids. At 560 km (348 miles) across, it should be large enough to form a perfect sphere, but Hubble Space Telecope photos reveal a misshapen southern hemisphere – the result of a massive impact early in its history. Uniquely among asteroids, Vesta is covered in bright, volcanic rock, suggesting it was once hot enough for geologial activity.

396
million km
from the Sun

253 Mathilde

Length: 66 km

Main belt asteroid

Mathilde has surprisingly weak gravity for its size (it is roughly 66 km or 41 miles in diameter). According to its effect on the NEAR probe during its 1997 flyby, it must have a density roughly equivalent to water. Its visible surface certainly seems to be fairly normal asteroidal rock, which suggests it must have a Swiss-cheese interior peppered with large voids.

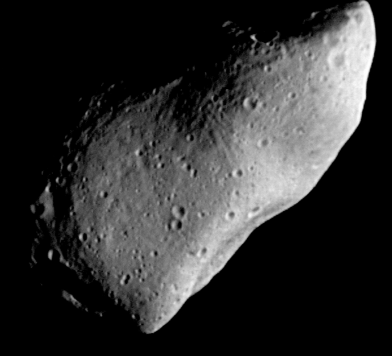

331

million km
from the Sun

951 Gaspra

Length: 18 km

Main belt asteroid

In 1991, before reaching Ida, the Galileo probe made a brief flyby of Gaspra. This small lump of rock, 20 km (12.5 miles) long, is an S-type asteroid, rich in silicate minerals, and also in metals. In the future, it may be possible to mine S-type asteroids – despite the difficulties, the purity of their metals might still make them more economical than extraction and processing on Earth.

Tempestuous Jupiter

The largest planet in the solar system, Jupiter's size and relative proximity to Earth meant that it was always a prime target for Earth-based observers even before the Space Age. The constantly shifting weather patterns could be tracked and recorded, even if they were poorly understood, while the motion of its satellites betrayed its approximate mass and density, revealing that it was a largely gaseous planet, rather than a solid ball of rock like the planets closer to the Sun.

However, Jupiter's location in the Solar System meant that exploration had to wait until a more powerful generation of launch vehicles had developed in the 1970s. The first probe to visit the giant planet was Pioneer 10, a relatively small and unsophisticated vehicle designed to carry out a quick reconnaissance of the innermost giant. It was soon followed by Pioneer 11, an identical spacecraft that used Jupiter's gravity to change its course and swing it on towards Saturn at greatly increased speed – a test run for the kind of orbital manoeuvres that are commonly used for spaceprobes today.

The Pioneers were a proving ground for technologies to be used on a more ambitious second pair of spacecraft – the missions that eventually became Voyagers 1 and 2. Launched in 1977, the Voyagers took advantage of a rare alignment among the outer planets that would allow a spacecraft to visit each of the giant planets in turn. Powered by the heat from a slowly decaying radioactive power cell, they carried a wide range of experiments and detectors, including far better cameras than those which had been available on the Pioneers.

Voyager 1 flew past Jupiter in March 1979, with Voyager 2 following in July. Their most impressive discoveries were the unique properties of the four giant Galilean satellites – Io, Europa, Ganymede and Callisto. When Voyager 1 turned its cameras to look back at Io after its close flyby, it photographed a huge plume of material arcing out from the moon's limb and high into the sky. Unexpectedly, Io was an extremely active world of volcanic eruptions, and further studies soon revealed enormous geysers of liquid sulphur blasting into

space, and large calderas of molten lava scattered across the planet's surface.

Europa proved even stranger – though it showed none of the surface activity seen on Io. Its curiously smooth, icy, and almost craterless surface betrayed the fact that it must be constantly resurfaced and wiped clean. Nevertheless, the surface is still criss-crossed with streaks of dirty pinkish ice. The Voyager scientists eventually concluded that Europa probably had a deep global ocean beneath a comparatively thin icy crust, and that the streaks were caused where brief cracks in the surface were healed by dirty water welling up from within. As with Io, Europa's interior must be kept warm by the constant flexing of its core under the influence of Jupiter's gravity.

Though considerably larger than the inner Galilean moons, the outer pair of Ganymede and Callisto proved something of a disappointment in terms of activity. Both appeared to be deep-frozen and inactive, though Ganymede at least showed a variety of light and dark terrains that suggested activity in the past. Callisto, in contrast, seemed to be unchanged since its formation – a heavily cratered ball of rock and ice.

With such a wealth of targets requiring further investigation, the decision to launch another probe to Jupiter was inevitable. While the Voyager probes could only take snapshots during a comparatively brief flyby, Galileo, which arrived at Jupiter in 1995 after a six-year voyage, was designed to go into orbit around the giant planet, circling within its satellite system for several years and executing a series of close encounters with all the Galilean moons. It even deployed a smaller atmospheric probe that parachuted into Jupiter's clouds and sent back a wealth of interesting data before it was inevitably crushed by the increasing pressure. Galileo eventually functioned for far longer than its intended two-year mission – ultimately, its power reserves dwindling, it was commanded to execute a death plunge into Jupiter's atmosphere in 2003. During eight years of operation, it completely transformed our view of the giant planet and its moons.

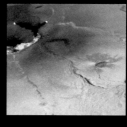

Top: The surface of Io seethes where exposed lakes of hot lava erupt from beneath the crust. Most are filled with the same molten sulphur compounds that cover Io's surface, and also help to colour the moon itself.

In early 2006, astronomers watched in fascination as a group of Jovian storms that had previously merged into a single superstorm gradually developed a colour to rival the Great Red Spot. It is thought that this new spot had grown deep and powerful enough to tap into the same deep layers of colourful chemicals as its larger sibling.

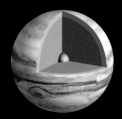

Beneath its colourful clouds, Jupiter is almost pure hydrogen. Increased pressure turns the atmosphere to liquid a little way into the planet, then transforms it to a liquid metal further down. At the centre lies a small core.

Distance from sun	Orbital period	Orbital eccentricity	Diameter	Cloudtop gravity	Rotation period	Axial tilt	Natural satellites
778.3	11.86	0.048	142,984	2.53	9.93	3.12	63+
million km	Earth Years		km	g	hours	degrees	

778.3

million km
from the Sun

Jupiter

Cassini portrait

Gas giant planet

The sheer size of Jupiter is staggering – the largest planet in the Solar System, it is so huge that any of the southern-hemisphere storms (the white spots and vortices on the left in this picture from the Cassini spaceprobe) would overwhelm the largest of Earth's continents. The Great Red Spot, meanwhile, is large enough to swallow the entire Earth twice over.

778.3

million km
from the Sun

Jupiter

Great Red Spot

Weather feature

The Great Red Spot is the most famous and long-lived of Jovian storms, observed since at least 1830, and perhaps since 1655. Its upper cloud layers tower 8 km (5 miles) above the surrounding atmosphere, but the root of the storm plunges deep into the chemical soup below, drawing up chemicals that condense to form red cloud in the chilly high altitudes at the top of the storm.

778.3

million km
from the Sun

Jupiter

Clouds and storms

Weather features

An image from Voyager 1 reveals the infinite detail of just a small stretch of Jovian cloud. Clouds of different colour, at different levels in the atmosphere, rotate around the planet at different rates. The currents where opposing regions meet can give rise to great loops and swirls called 'festoons', which occasionally detach themselves and take on an independent life as storms.

778.3

million km
from the Sun

Jupiter

Infrared Jupiter

Gas giant planet

An infrared image of Jupiter, captured by the Earthbound Gemini telescope, shows that the planet is pumping out tremendous amounts of heat. In fact, Jupiter releases considerably more heat than it receives from the Sun, suggesting its clouds conceal a powerful energy source. It is thought Jupiter's interior is still slowly collapsing and compressing, releasing heat in the process.

Jupiter

Belts and zones

Weather features

Jupiter's dark and light bands of cloud are known as belts and zones respectively. The lighter zones are higher in the atmosphere than the dark belts. The bands are similar to high and low-pressure regions in Earth's weather systems, but the strong 'coriolis forces' generated by Jupiter's rapid rotation stretch them out parallel to the planet's equator.

778.3

million km
from the Sun

Jupiter

Jovian rings

Ring system

Like all the giant planets, Jupiter is surrounded by a system of rings. However, the Jovian rings are somewhat feeble, barely more than a thin plane of dust around the inner moons that only becomes visible when backlit by the Sun. For this reason, the rings were not discovered until Voyager 1 looked back to photograph the night side of Jupiter after its 1979 rendezvous.

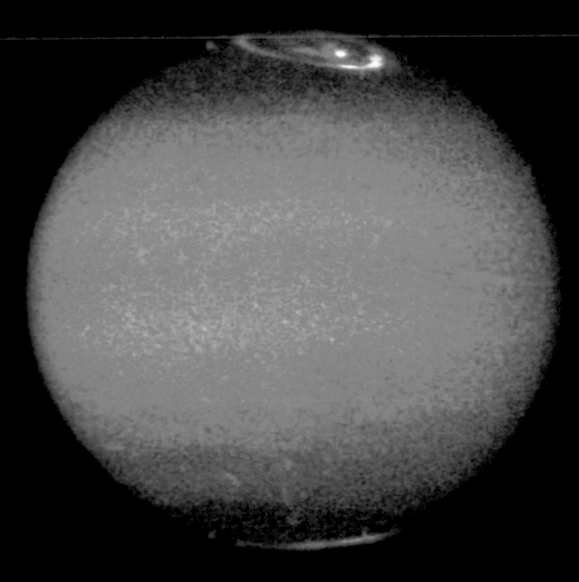

778.3

million km
from the Sun

Jupiter

Jovian aurorae

Atmospheric feature

Jupiter's magnetic field is over ten times stronger than Earth's – its influence is felt all the way to the orbit of Saturn. The magnetism is driven by currents in the liquid metallic hydrogen of the planet's mantle, and just as on Earth, it sweeps up charged particles from the solar wind and directs them down onto the magnetic poles, creating brilliant aurorae as they plunge into the atmosphere.

778.3

Jupiter

Comet bruises

million km
from the Sun

Atmospheric feature

In 1994, astronomers watched as fragments of Shoemaker-Levy 9, a comet previously torn to shreds by a close encounter with Jupiter's gravity, plunged into the giant planet, setting off the largest explosions ever seen in the Solar System. Jupiter is thought to act as a 'guardian angel' for the inner Solar System, disrupting or destroying many comets that might otherwise harm Earth.

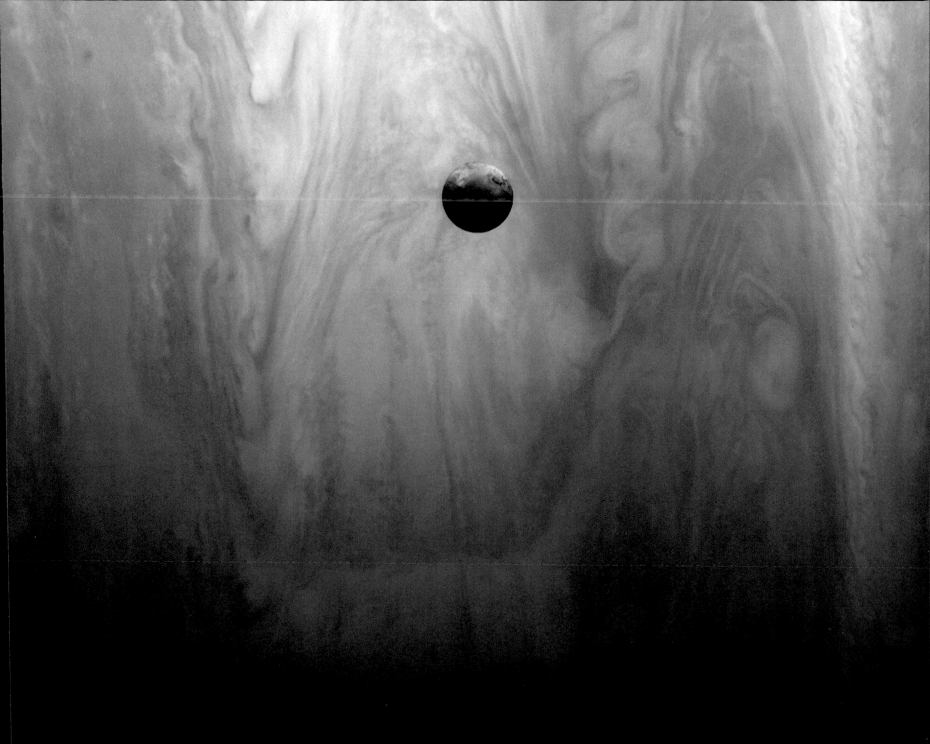

421.6

thousand km
from Jupiter

Io

Io with Jupiter

Rocky natural satellite

Jupiter's innermost major moon, Io, hangs on the planet's 'twilight zone' in this image, taken from the passing Cassini probe. Io orbits Jupiter in just over 42 hours, and its slightly elliptical orbit causes the effect of Jupiter's gravity to vary. This causes the entire satellite to flex slightly, creating a tidal effect far more powerful than the tides between the Earth and Moon.

421.6

thousand km
from Jupiter

Io

Diameter: 3,643 km

Rocky natural satellite

The enormous tides Jupiter exerts on Io help to keep the moon's interior hot and active – as a result, Io is the most volcanically active world in the entire Solar System. Images such as this Galileo portrait allowed astronomers to identify more than 200 volcanic craters or calderas over 20 km (12.5 miles) in diameter. Most of Io's volcanoes erupt colourful sulphur compounds.

421.6

thousand km
from Jupiter

Io

Sulphur eruption

Volcanic plume

The most obvious volcanic features on Io are the mushroom-shaped plumes of material that burst from its surface and arc hundreds of kilometres above the moon. The plumes are geysers similar to the gushers of steam and water found in volcanic regions of Earth, only here they are fuelled by molten sulphur dioxide. Most material from the plumes falls back onto Io as 'snow'.

670.9

thousand km
from Jupiter

Europa

Diameter: 3,122 km

Rock/ice natural satellite

The second, and smallest, Galilean moon is Europa. Very different from its neighbours, it is a pinkish-white world with a very thin oxygen atmosphere. The crust is actually largely made of water ice, and is criss-crossed by countless pinkish scars. Craters are few and far between, however, which suggests that the moon is consistently resurfaced, wiping away their traces.

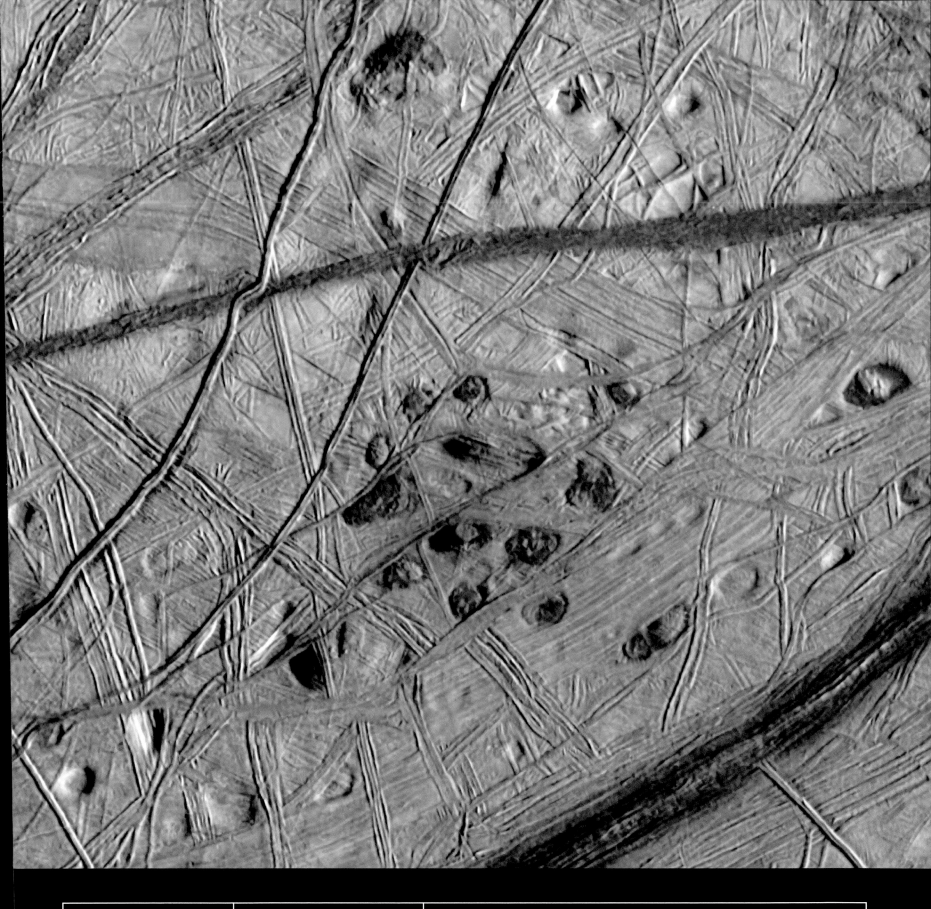

thousand km
from Jupiter

Europa

Freckles and ridges

Ice features

Closer in to Europa, the icy nature of the surface is clearer. Images such as this one from Galileo resolve the moon's scars into closely spaced parallel lines. Despite all this surface detail, Europa is cue-ball smooth – if it were blown up to the size of the Earth, it would have no features on it higher than about 200m (660 ft). This is because the shifting ice flattens out high or low features.

670.9

thousand km
from Jupiter

Europa

Double ridges

Ice features

A colour-enhanced close-up of Europa's scars reveals that, although they are made of ice, they are contaminated with dirty brown, sulphurous chemicals. It's now clear that Europa's icy crust floats on top of a subsurface ocean, kept liquid by the same tidal heating that affects Io. When the surface occasionally cracks apart, dirty water from below wells up and freezes, healing the gap.

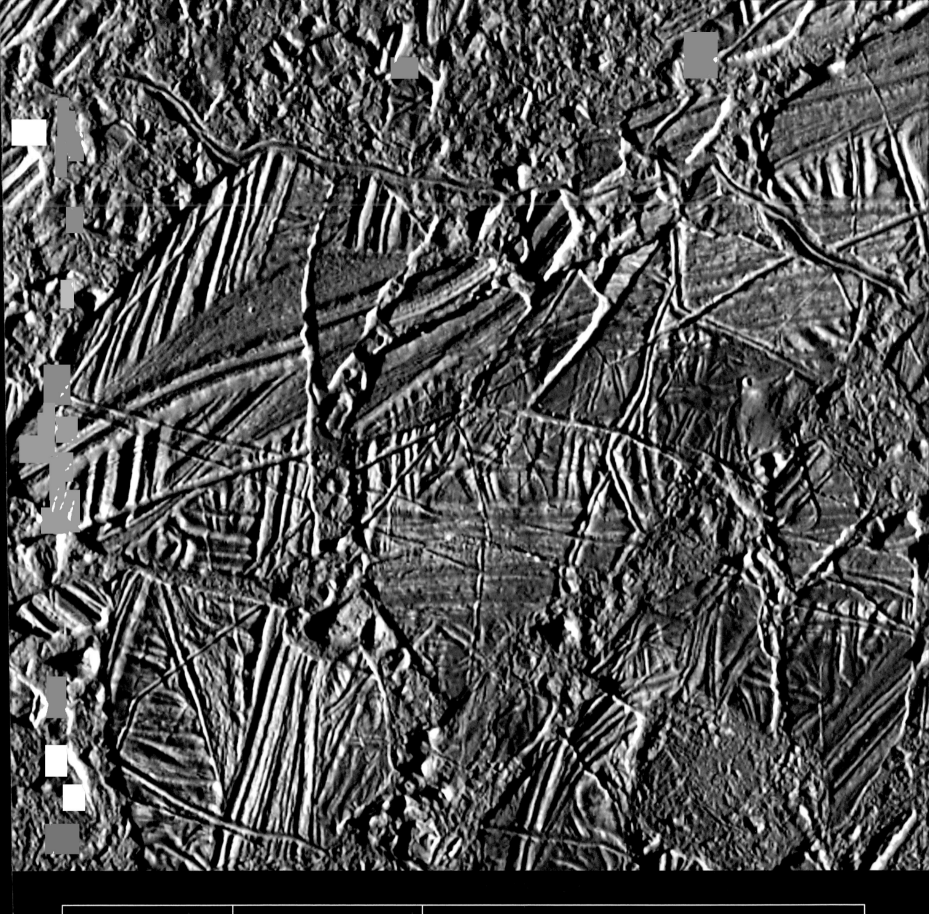

670.9

thousand km
from Jupiter

Europa

Conamara Chaos

Ice feature

Other areas of Europa's crust resemble the pack ice of Earth's own polar regions. However, the ice on Europa is a far more formidable layer – it is thought to form a cap 10 km (6 miles) thick on top of an ocean perhaps 100 km (60 miles) deep. One of the great mysteries of the Solar System is whether life might thrive around undersea volcanoes on Europa as it does on Earth.

1.07

million km
from Jupiter

Ganymede

Diameter: 5,262 km

Rock/ice natural satellite

Ganymede is a world of whites and browns, the largest satellite in the Solar System, and third of the Galilean moons in order from Jupiter. Although it is not so immediately striking as Io or Europa, the clear division of the surface into comparatively old, dark areas and brighter strips between them shows that Ganymede has had an active past.

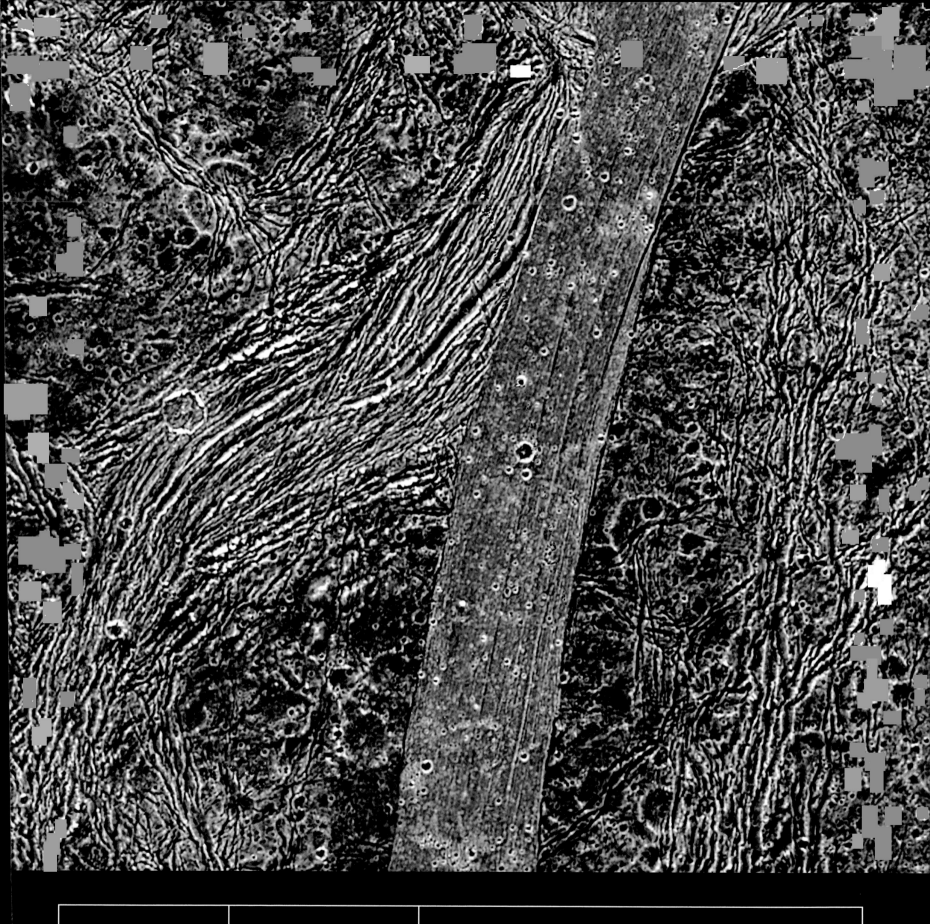

1.07

million km
from Jupiter

Ganymede

Arbela Sulcus

Ice feature

Long strips of parallel ridges called sulci show where Ganymede's surface has pulled apart, causing strips of terrain to slip over so they now overlap each other like a line of toppled dominoes. Fresh material welling up from below the crust then freezes, helping to seal any gaps. In some sulci, it's possible to trace the broken outlines of older features such as craters among the ridges.

1.883

million km
from Jupiter

Callisto

Diameter: 4,821 km

Rock/ice natural satellite

Compared to the other Galilean moons, outermost Callisto has changed little since its formation. Combined with its large size and position in the 'line of fire' towards Jupiter, this has made it possibly the most heavily cratered world in the Solar System. Impacts deep enough to penetrate the dark outer crust have scattered fresh icy ejecta across the satellite's surface.

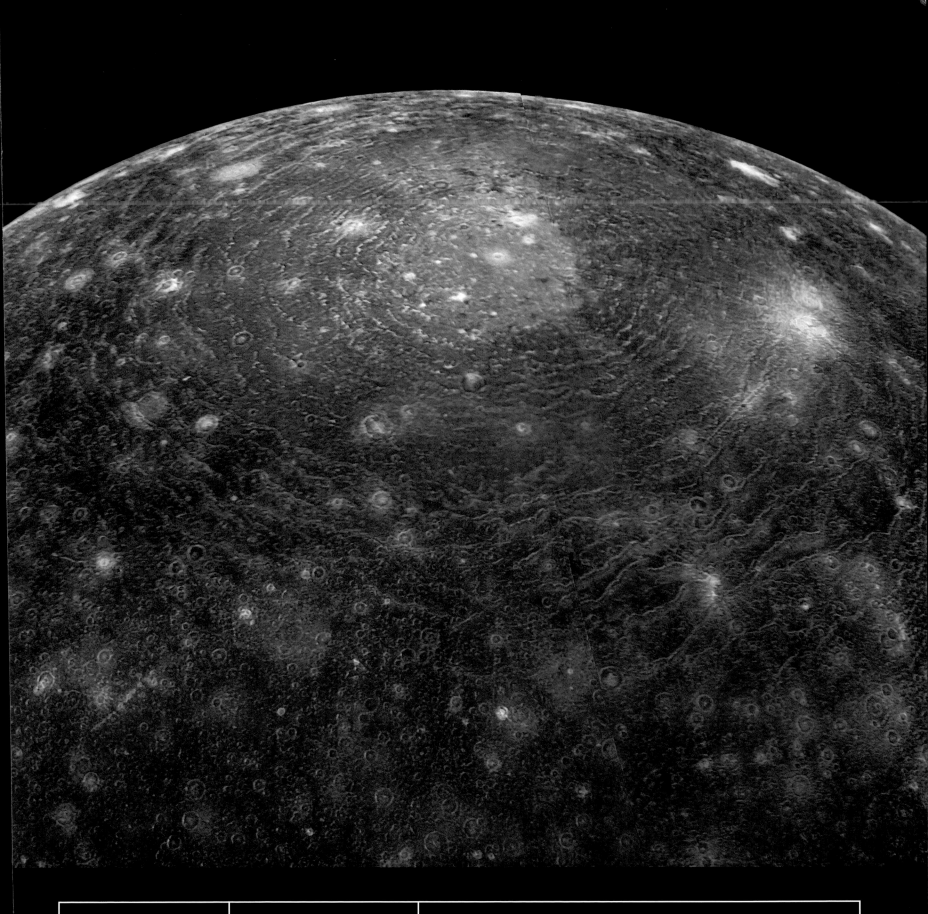

 **million km**
from Jupiter

Callisto

Valhalla Basin

Impact basin

Palimpsests are large, ancient impact basins with bright centres found on both Ganymede and Callisto. It seems that these major impacts punctured the upper crust and allowed fresh, slushy ice to well up from within the moons, wiping away all traces of earlier activity. On Callisto, the palimpsest at the heart of the Valhalla impact basin is by far the largest.

Circling Saturn

The famously ringed planet Saturn marked the outer limit of the Solar System throughout until well into the telescopic era. When Galileo first turned his primitive telescope on it, he could see that there was something wrong with its shape, and ultimately concluded that the planet was orbited by two very large, very close moons. Others believed that Saturn itself was misshapen, or that it had 'handles'. The situation was thrown into further confusion in 1612, when the rings did one of their periodic 'disappearing acts' as their own narrow plane lined up precisely with the direction of Earth. It was not until 1655 that Dutch astronomer and instrument-maker Christiaan Huygens, using a far more powerful telescope than Galileo's, correctly concluded that the planet was surrounded by a ring with enormous extent but very little thickness. And it took another two centuries for James Clerk Maxwell to correctly explain that the rings must be made of countless small chunks of material, each in its own perfect orbit around Saturn. By that time, other astronomers had identified numerous different rings and relatively empty divisions between them.

As a planet, though, Saturn remained obscure – an almost featureless yellowish-white disc on which occasional bright white storms appeared. It was clear that the planet was a gaseous giant like Jupiter, but little else was known about it.

Things began to change in 1979, when Pioneer 11 made the first flyby of the Saturnian system – for example, measurements of Saturn's mass showed that it was the least dense planet in the Solar System (in theory, it would float in water). It is also the most distorted of all the planets – its low density and fast 10.2-hour rotation combine to make the planet bulge significantly around its equator, so the diameter from pole to pole is considerably less than the equatorial diameter.

Our exploration of the ringed planet and its huge system of moons began to accelerate with the arrival of the Voyager probes in 1980 and 1981. For the first time, the extent of activity in Saturn's atmospheres became clear – this muted planet, it seems, is just as turbulent as Jupiter, but the cooler temperatures of its upper atmosphere allow a hazy layer of ammonia clouds to crystallize, blurring and bleaching the more colourful clouds below.

The rings, meanwhile, proved even more spectacular than was previously suspected – each vast halo was made up of hundreds of separate, sharply defined ringlets, occasionally disrupted by the gravitational influence of larger moons and moonlets.

Saturn's system of moons is very different from that around Jupiter. While both worlds have several dozen satellites, most of which are captured asteroids plucked from nearby space as they were passing, the innermost groups of 'natural' satellites (which formed in orbit around their parent planet) contrast strongly. The Jovian system is dominated by the four giant Galilean moons, which dwarf the planet's other satellites. Saturn has just one such giant moon, Titan, but a host of medium-sized satellites made of rock and ice in varying mixtures. Titan was a particular target for the Voyager missions, since it is the only moon in the Solar System with a substantial atmosphere. However, Voyager 1's close flyby revealed little more than a uniform, opaque orange haze, a mixture of nitrogen and methane.

Like Jupiter, Saturn was crying out for a follow-up mission able to orbit the planet for several years. After a decade's gestation and a seven-year flight, the sophisticated Cassini spaceprobe finally entered orbit around Saturn in 2004. Equipped with high-resolution cameras, infrared detectors capable of seeing through Titan's atmosphere, and even an automated lander destined for Titan, the ongoing Cassini mission has transformed our view of Saturn and its moons once again, and is still sending back valuable data and spectacular images.

When the Huygens probe touched down on Saturn's giant moon Titan in 2005, it landed in the estuary of a methane river, surrounded by icy rocks. At Titan's low temperatures, methane exists as ice, liquid and vapour, behaving in the same way that water does on Earth.

This Cassini image captures a jet of water vapour shooting out from the surface of Enceladus. Liquid water just below the surface of such a small moon contradicts a lot of established wisdom about geological activity on small worlds.

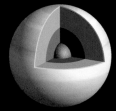

Beneath Saturn's outer clouds, hydrogen and helium rapidly turn to liquid, while at greater depths hydrogen molecules split up to form an ocean of liquid metallic hydrogen. A small core of rock and ice lurks at the centre.

Distance from sun	Orbital period	Orbital eccentricity	Diameter	Cloudtop gravity	Rotation period	Axial tilt	Natural satellites
1.43	29.46	0.056	120,536	1.07	10.66	26.73	561
billion km	Earth years		km	g	hours	degrees	

1.43

billion km
from the Sun

Saturn

Cloud belts

Weather features

Although at first glance Saturn appears a placid world compared to Jupiter, appearances are deceptive. This colour-enhanced image from the Cassini orbiter shows that a great deal of activity is going on, though it is usually veiled by Saturn's outer haze of ammonia. From Earth, only occasional 'white spots' reveal that Saturn is a turbulent, stormy world.

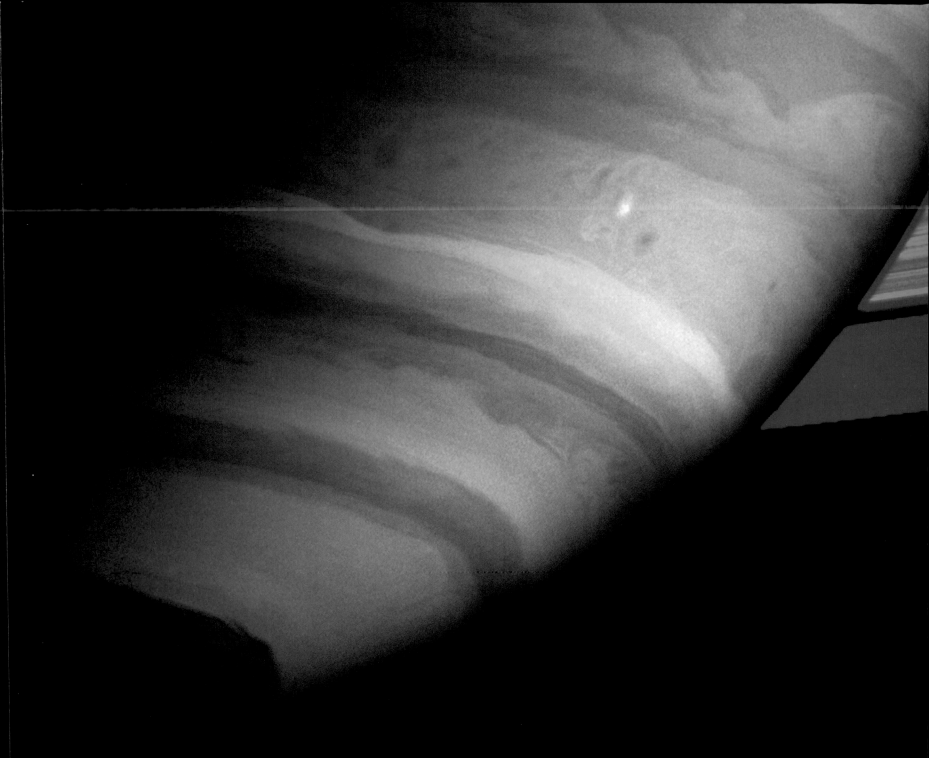

1.43

billion km
from the Sun

Saturn

Dragon storm

Weather feature

Like Jupiter, Saturn is girdled by cloud belts and zones parallel to its equator. The pinkish-red swirl near the top of this Cassini photo is a vast electrical storm in Saturn's atmosphere, large enough to engulf the United States and producing lightning bolts 1,000 times more powerful than those on Earth. Such storms are a more or less constant feature in the atmospheres of both Jupiter and Saturn.

1.43
billion km
from the Sun

Saturn

The major rings

Ring system

This spectacular panorama reveals the fine structure of Saturn's rings, like grooves in a vinyl record. Gaps between the rings, and differences in their particles, define separate regions – from Saturn out, the main ones are the diaphanous D ring, the pale C ring, the bright B ring, the almost-empty Cassini Division, the A ring (including the narrow Encke division), and the thread-like F ring.

1.43

billion km
from the Sun

Saturn

Ring shadows

Ring system

The rings cast deep shadows onto the globe of the planet itself, creating complex patterns of striation. Individual ring particles generally all follow perfectly circular orbits above Saturn's equator, in order to avoid collisions with one another, but their perfect appearance can be disturbed by the gravity of Saturn's inner moons, creating short-lived 'spokes' and 'ripples' in the rings.

Saturn

Edge-on rings

Ring system

From directly above Saturn's equator, the rings themselves are a vanishingly narrow line, just a few hundred metres across, with the tiny moon Enceladus hovering just above them. With Saturn's northern hemisphere tilted away from the Sun, sunlight is taking a longer path through the atmosphere and is being scattered by gas as it travels, creating the distinctive sky blue tint.

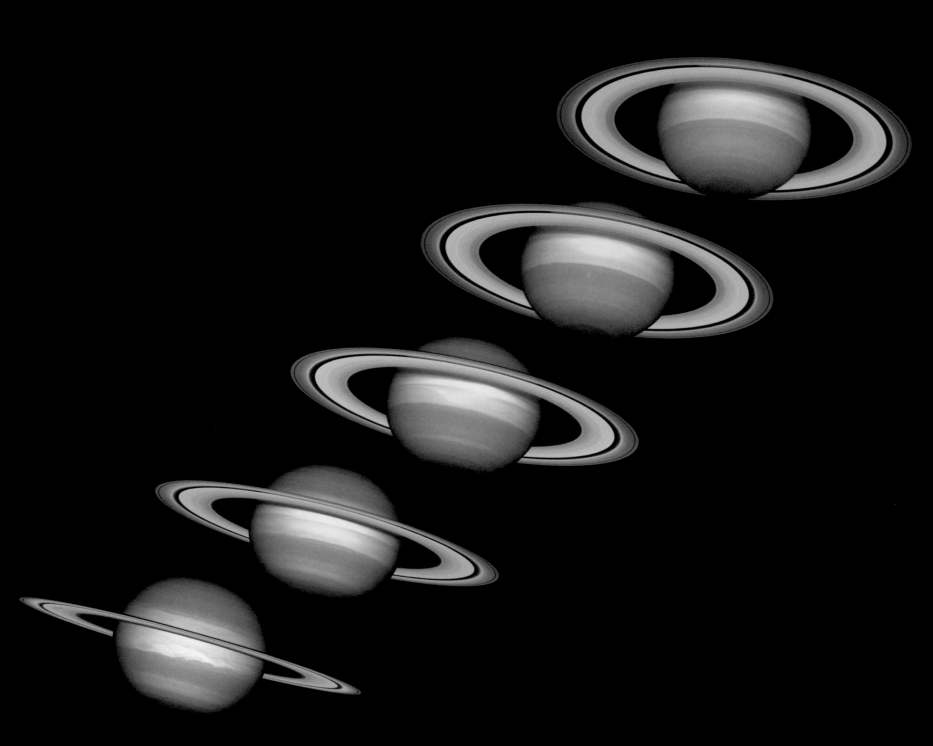

1.43

billion km
from the Sun

Saturn

Tilting rings

Ring system

Because Saturn is tilted at an angle of 26.7 degrees from 'straight up', it displays different aspects to Earth throughout its orbit. Every 15 years, Saturn lies 'edge-on' to Earth, and the rings disappear from view completely. This sequence of images from the Hubble Space Telescope traces the rings as they 'open up' after one of these disappearances to reveal their southern side.

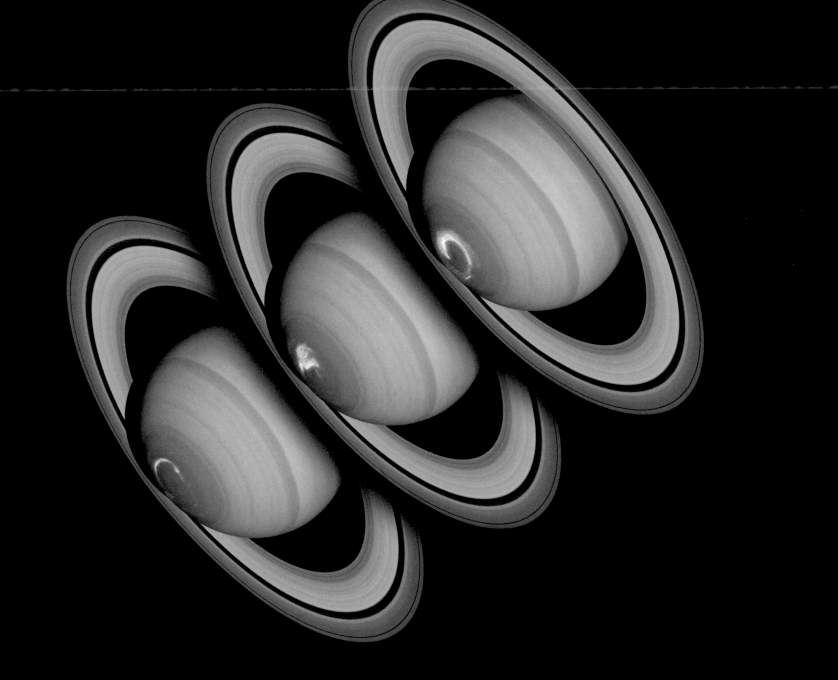

1.43

billion km
from the Sun

Saturn

Aurorae

Atmospheric feature

Saturn's aurorae, formed as the solar wind of particles from the Sun is funnelled into the planet's own magnetic field, form an unusual spiral pattern not seen on either Earth or Jupiter. This image shows their ultraviolet radiation in blue – without this processing the aurorae would appear red, and their light would probably be washed out by the brightness of Saturn's disc.

139.4

thousand km
from Saturn

Prometheus

Length: 119 km

Icy natural satellite

Numerous tiny moonlets orbit among the main rings, but the innermost objects that have been well photographed are Prometheus and Pandora. These 'shepherd moons' orbit just inside and outside the thread-like F ring, helping to keep its particles confined on their narrow, close-packed path. Here Prometheus is in the background, with its outer neighbour Janus in the foreground.

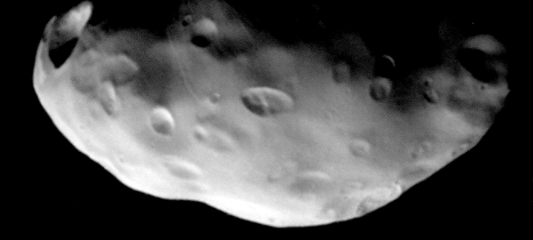

Pandora

Length: 103 km

Icy natural satellite

The second of the F ring's shepherd moons is Pandora, an irregularly shaped, cratered world 84 km (52 miles long). These moons may be fragments left over from a larger satellite that broke up to form the rings themselves. Their cratered surfaces suggest they are subjected to occasional impacts that fling off material to help replenish the rings.

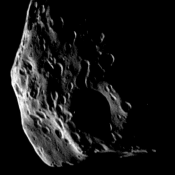

151.4

thousand km
from Saturn

Epimetheus

Diameter: 138 km

Icy natural satellite

Sharply-chiselled Epimetheus looks like a broken remnant of some larger body, and it probably shares an origin with Janus. Today, it has a complex relationship with its sibling: usually one moon orbits about 50 km (30 miles) closer to Saturn than the other, meaning that it travels faster and eventually catches the other one up. When this happens, the two moons 'swap orbits'.

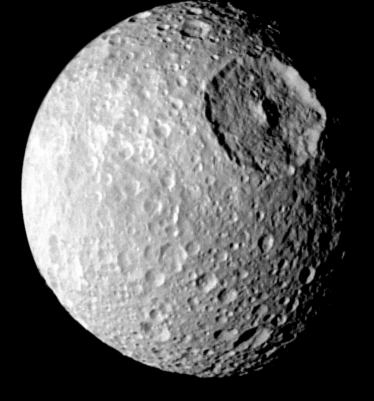

185.5

thousand km
from Saturn

Mimas

Diameter: 397 km

Icy natural satellite

In a case of life imitating art, Mimas bears an impressive resemblance to the Death Star space station from the Star Wars movies. The 'radar dish' feature is in fact the huge Herschel Crater, at 140 km (88 miles) across, more than one-third the diameter of the entire moon. It is about as big as a crater can get without shattering a moon completely.

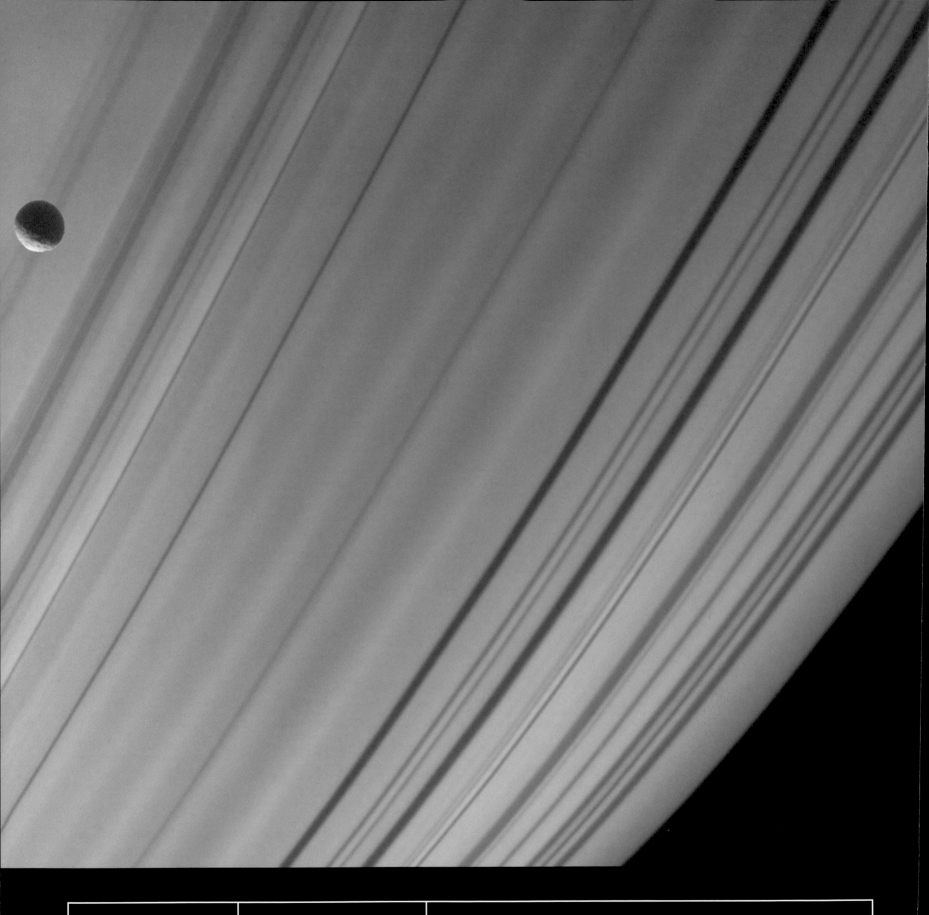

185.5

thousand km
from Saturn

Mimas

Mimas and ring shadows

Icy natural satellite

Mimas is the innermost of the major Saturnian moons, some 397 km (247 miles) across and orbiting within the sparse, broad E ring. It can frequently be seen against a backdrop of rings and Saturn itself as in this spectacular image, where Mimas sits in front of Saturn's blue-tinted northern hemisphere. Ring shadows across the globe of Saturn create a curtain-like effect.

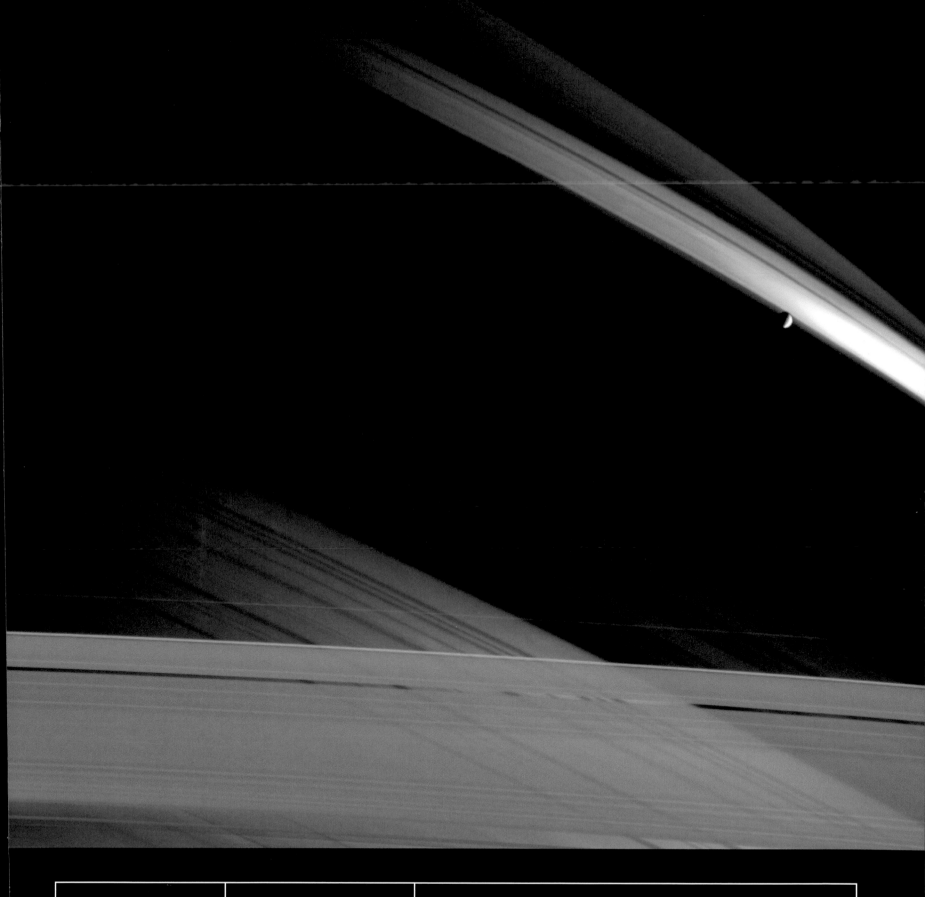

185.5

Mimas

Mimas above the A ring

Icy natural satellite

In another image from the Cassini probe, tiny Mimas again sits in front of Saturn's own globe, with dark ring shadows behind it. The bright swathe of Saturn just above Mimas is a region where sunlight is shining through the Cassini Division. The A ring runs across the bottom of the picture, with the Cassini Divison itself beneath it, and the narrow F ring above.

thousand km
from Saturn

Enceladus

Diameter: 512 km

Icy natural satellite

Saturn's second major moon, Enceladus, has the most reflective surface of any world in the Solar System, an almost pure snowy-white (this image has been processed to emphasize colour differences). An icy composition and tidal heat generated by its proximity to Saturn helps create geysers of water vapour that periodically erupt into space and fall back to the ground as snow.

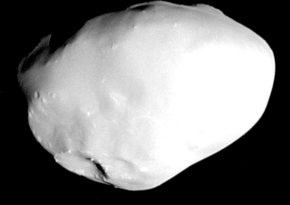

294.7

thousand km
from Saturn

Telesto

Length: 30 km

Icy natural satellite

Tiny Telesto is a rare example of a 'co-orbital' satellite. It shares an orbit with the much larger Tethys, but moves around Saturn a constant 60 degrees ahead of its big sister, keeping well out of its way. Another small moon, Calypso, trails a similar distance behind Tethys. These co-orbital moons are the satellite equivalent of the Trojan asteroids that share the orbit of Jupiter.

294.7

thousand km
from Saturn

Tethys

Diameter: 1,072 km

Icy natural satellite

The third major moon, Tethys, is considerably larger than either Mimas or Enceladus. Its surface is relatively bright, yet it has many more craters than Enceladus. The largest one visible in this Cassini image is named after Penelope, Odysseus's wife – appropriately, a much larger crater called Odysseus dominates the moon's other face.

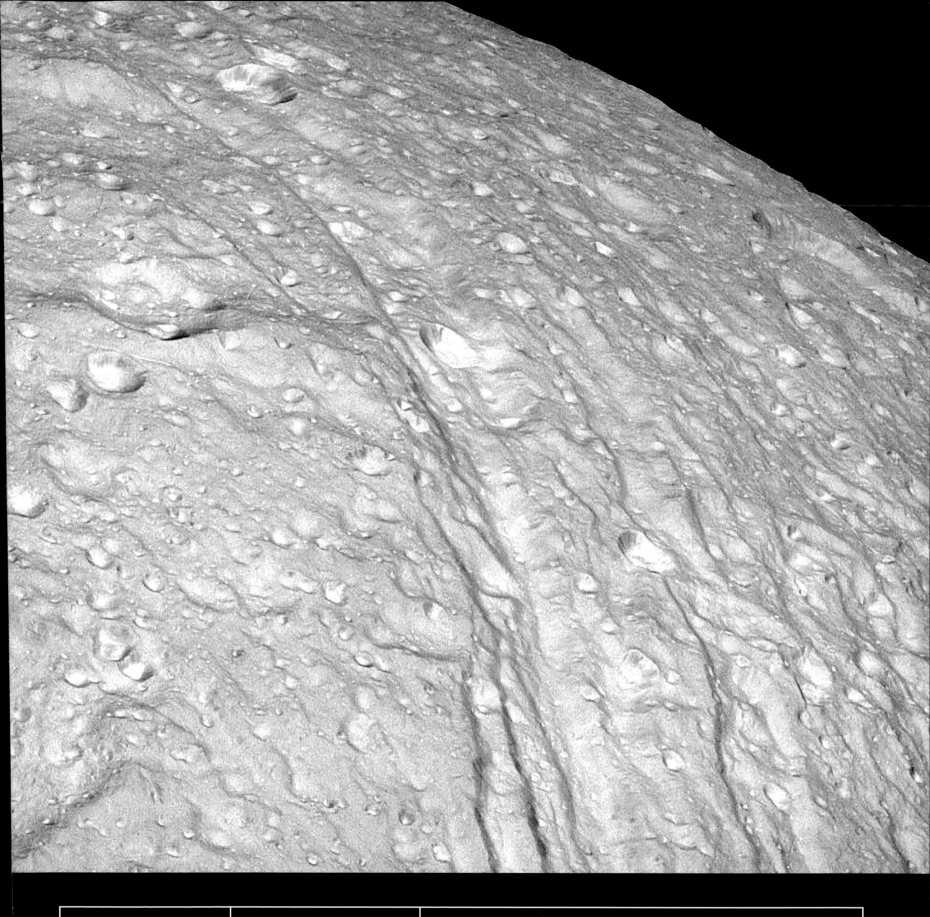

294.7

thousand km
from Saturn

Tethys

Ithaca Chasma

Crustal fault

Ithaca Chasma is a long trench or valley system that wraps itself around much of the planet. Although it is some distance from the huge Odysseus crater, it runs parallel to the crater wall. The crater itself is unusually shallow for its size, and it's thought that the icy material in Tethys's crust slumped under its own weight, flattening the crater but opening this huge fault line in the process.

377.4

Dione

Diameter: 1,120 km

Icy natural satellite

The electronic cameras on board spaceprobes such as Cassini are highly sensitive to variations in light levels, but are not built to detect colour. In order to reconstruct a natural-colour image such as this one of Dione hanging above Saturn's ring plane, separate images are taken through red, green and blue filters, and then combined electronically back on Earth.

377.4

thousand km
from Saturn

Dione

Wispy terrain

Ice feature

When the Voyager probes first photographed Dione in the early 1980s, the most prominent features were a network of bright streaks on a dark background, soon nicknamed 'wispy terrain'. (Some of these streaks can be seen on the moon's left limb in this Cassini image.) Much closer flybys from the Cassini orbiter have revealed that these streaks are in fact long, bright ice cliffs.

527

thousand km
from Saturn

Rhea

Diameter: 1,528 km

Icy natural satellite

Rhea is far more heavily cratered and darker than Tethys or Dione, suggesting that it has changed little since its formation. This contradicts the idea that larger bodies will be more geologically active due to the heat they build up during their formation. The best explanation is that Rhea's higher density made it freeze completely solid in a way that its inner neighbours did not.

1.43

billion km
from the Sun

Saturn

Dione, Tethys and Pandora

Icy natural satellites

This beautiful Cassini image shows Dione, Tethys and tiny Pandora from just above Saturn's ring plane. The immense distances and high magnifications at which Cassini and other spaceprobes sometimes take their pictures can be deceptive – in this tableau, Dione and Pandora are actually on the near side of the rings, while Tethys is on the far side.

1.22

million km
from Saturn

Titan

Diameter: 5,150 km

Icy natural satellite

Titan is the largest of Saturn's moons by a long margin – it sits in the middle of the Saturnian system flanked by satellites of dwindling size to either side. At one time, it was thought to be the largest satellite in the entire Solar System. But astronomers were deceived by the thick atmosphere, on display in this Cassini image – in fact it's second in size after Jupiter's moon Ganymede.

1.22

million km
from Saturn

Titan

Haze

Atmospheric feature

Titan is unique among moons in its possession of a substantial atmosphere. This envelope of gases is dominated by nitrogen, like Earth's atmosphere, but contains two per cent methane and no free oxygen. This bright orange, opaque veil turned the moon into an impenetrable mystery for early spaceprobes and blocks 90 per cent of sunlight from ever reaching the surface.

| 1.22 | *Titan* | The Cassini probe, which entered orbit around Saturn in 2004, went equipped to unveil the mysteries of Titan. Its infrared cameras were able to photograph the planet using radiations where Titan's atmosphere becomes transparent. The images they revealed bear a startling resemblance to Earth's own pattern of continents and oceans. |
| **million km** from Saturn | Titan unveiled

Icy natural satellite | |

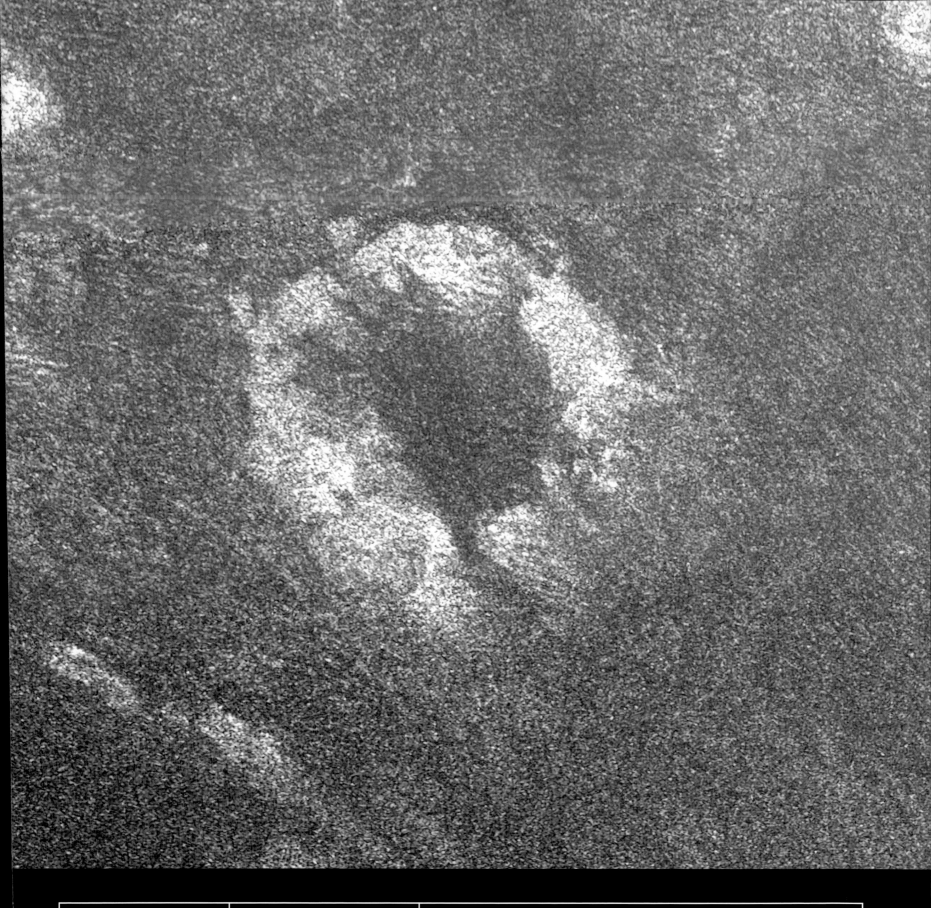

1.22

million km
from Saturn

Titan

Guabonito

Cryovolcanic feature

Methane breaks down rapidly in a planet's atmosphere, so for Titan to have such a methane-rich atmosphere today, it must be continually replenished. Some scientists believe that this methane comes from 'cryovolcanic' eruptions of icy mud. It's possible that the Guabonito structure in Titan's Xanadu region is the caldera of a cryovolcano, but it might also be an ancient impact crater.

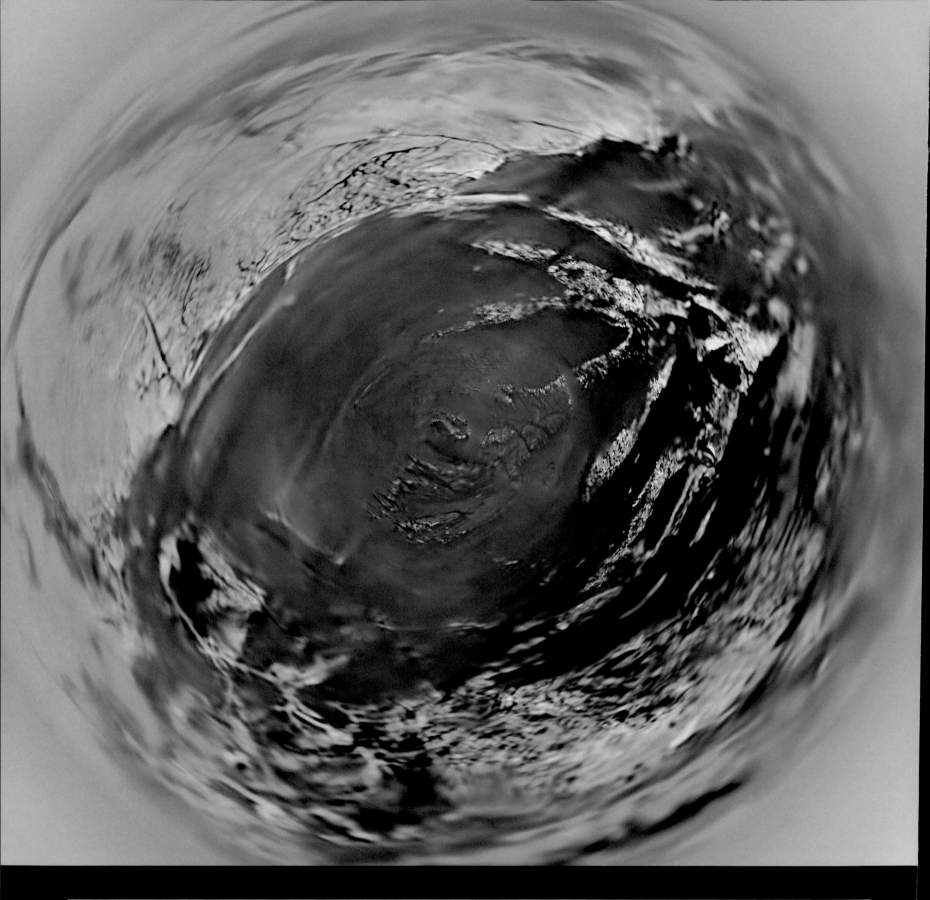

1.22

million km
from Saturn

Titan

Titan's surface

Erosion features

Cassini carried a lander called Huygens with it to Saturn, designed to make a controlled descent to Titan and send back data from the surface. As the clouds parted, they revealed another eerily Earth-like view, seen here in a fish-eye projection. Traces of coast-like erosion are everywhere, bearing out the theory that Titan has a 'methane cycle' that parallels Earth's own 'water cycle'.

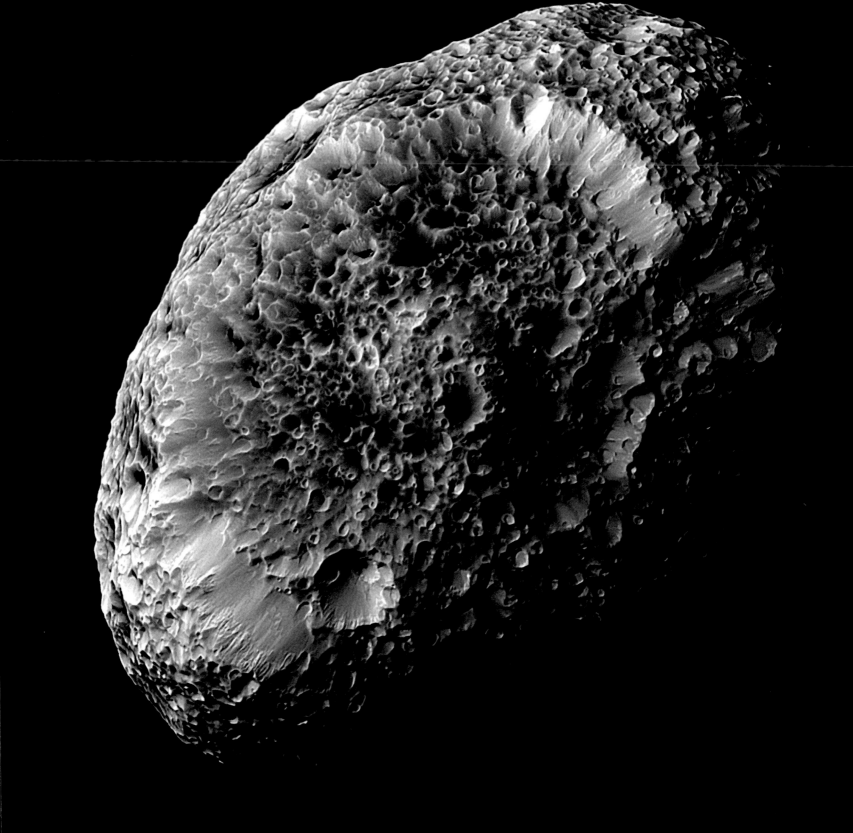

1.48

million km
from Saturn

Hyperion

Length: 370 km

Icy natural satellite

Beyond Titan lies the bizarre and beautiful moon Hyperion. Its irregular shape and unpredictable rotation indicate that it is the vestige of a larger moon that once suffered a catastrophic collision. However, Hyperion's strangely sponge-like surface, which in places resembles an impact frozen in 'mid-splash', still largely defies explanation.

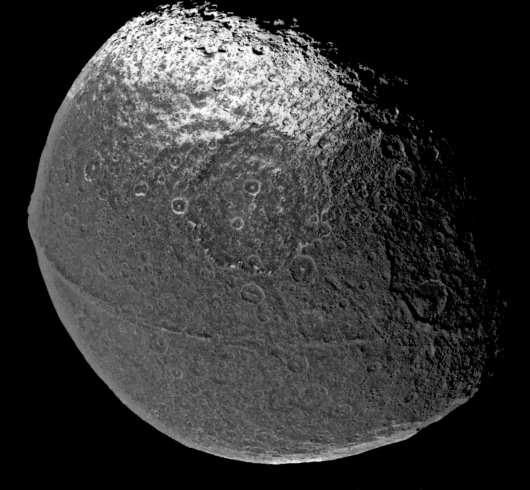

3.56
million km
from Saturn

Iapetus

Diameter: 1,436 km

Icy natural satellite

The outermost major moon of Saturn, Iapetus has its own unique claim to fame – it is a world of two halves. Locked with one face permanently towards Saturn, the side of the moon that faces 'along' its orbit has a coating of dark material that makes it far fainter than the uncoated, trailing hemisphere. Another strange feature is the long straight ridge running around the equator.

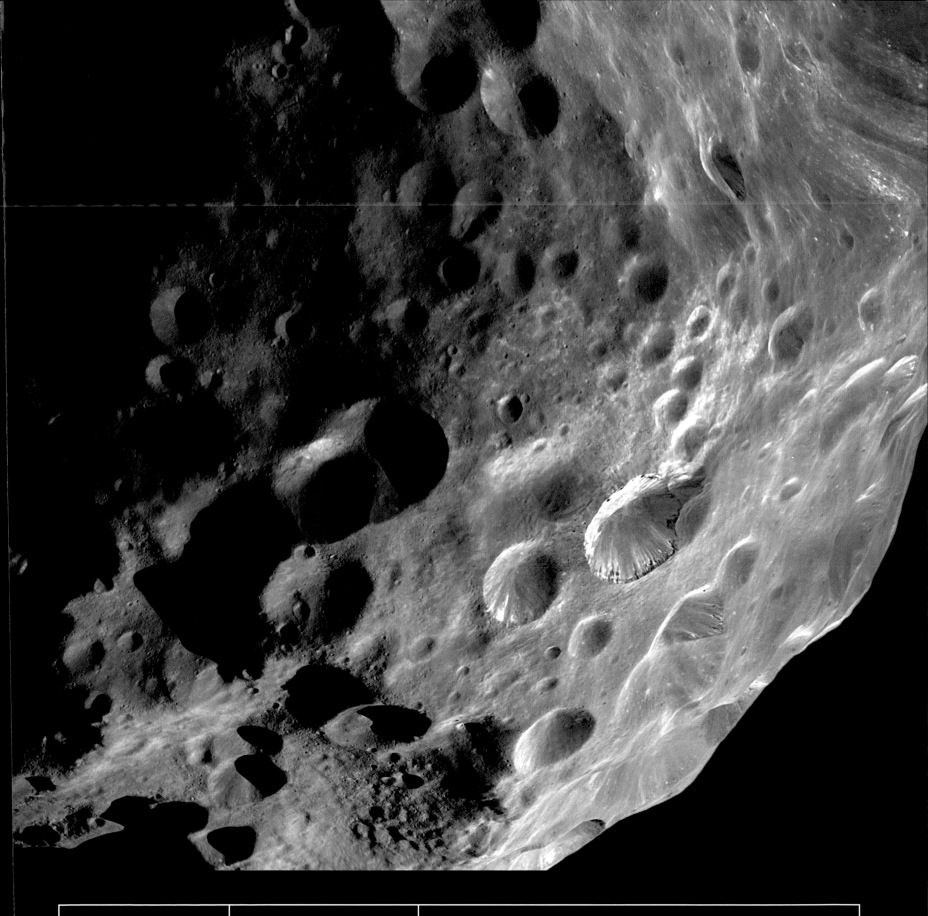

12.95

million km
from Saturn

Phoebe

Length: 230 km

Icy captured satellite

The largest and innermost of Saturn's 'captured' moons, Phoebe is almost certainly a centaur – an icy asteroid pulled into an orbit the 'wrong way' around Saturn. The best explanation for the strange dark coating on Iapetus is that it's a result of dark material being stripped from Phoebe and falling towards Saturn, only to be swept up by its inward neighbour.

Tilted Uranus

The first new planet of the telescopic era, Uranus has always been something of a mystery world. Its discoverer, William Herschel, at first mistook it for a comet, before its slow track across the sky revealed that it must in fact be much larger and much further away. Until recently, even the most advanced Earth-based telescopes showed it only as a blurred blue-green disc, roughly halfway in size between the terrestrial planets and the true giants such as Jupiter and Saturn.

One thing that was clear, however, was Uranus's bizarre pattern of seasons. The planet's four largest moons, Ariel, Umbriel, Titania and Oberon, were discovered in the late 18th and 19th century, and when astronomers traced their orbits, it was clear that something strange was happening – instead of moving from one side of the planet to the other and back again, these satellites were looping 'above' and 'below' the planet. Since, almost without exception, moons orbit above a planet's equator, this was clear evidence that the planet's equator was tilted so far over that it was almost upright. In fact, it turned out that Uranus's entire axis of rotation is tilted over at 98 degrees from the vertical – so the planet's north pole points slightly 'downwards', below the plane of its orbit.

The strange tilt has a huge effect on the planet's seasons. While a planet like Earth (tilted by about 23 degrees from 'upright') has relatively mild seasons as first one hemisphere and then the other is pointed more towards the Sun and receives slightly more solar energy, on Uranus, summer is a period of eternal daylight, while winter is a time of perpetual darkness. As the planet completes its 84-year orbit around the Sun, many areas of the planet experience a 'day' that lasts for decades, a few years in which the sun rises and sets with Uranus's 17-hour rotation period, and then a decades-long 'night'.

The effect this has on the Uranian weather became clear when Voyager 2, the first and so far

only spaceprobe to visit the planet, flew past it in 1986. Compared to the stormy turbulence of Jupiter and Saturn, Uranus was something of a disappointment, a featureless pale turquoise orb. It seemed that strong currents moving warm gas from the summer to the winter hemisphere completely disrupted any other weather patterns that might try to form.

Fortunately, the Voyager 2 encounter offered other highlights. It revealed a system of planetary rings very different from those around Saturn, and an impressive family of satellites. The rings had been discovered in 1977, when Earthbound astronomers watching Uranus eclipse a distant star noticed that the star 'flickered' several times before and after the main eclipse. This suggested that the planet was surrounded by a system of narrow, well-defined rings (eleven of these are now known).

Among the major moons of Uranus, the most impressive was undoubtedly the smallest. Miranda was a late discovery, found by Dutch astronomer Gerard Kuiper in 1948. It orbits much closer to Uranus than the other four large moons, and is just 480 km (300 miles) across, but Voyager 2 was able to fly quite close to it, recording a puzzling, jumbled landscape with aspects that astronomers still struggle to explain.

The years that have passed since Voyager 2's flyby have seen major advances in Earth-based telescopes – not least the success of the orbiting Hubble Space Telescope – and this has helped to make up for a lack of new missions to the outer giants. When the telescope turned its gaze towards Uranus in 1997, it revealed a changed world, suddenly alive with several bright and active storms, and distinct banded weather systems. The most likely explanation for this sudden change is the onset of Uranian spring, as the planet's axis of rotation tilts away from the Sun and large areas of the planet experience a 'normal' cycle of day and night.

The rings of Uranus are far darker than those around Saturn, and are confined to eleven narrow threads, of which the brightest (on the left in this image) is the Epsilon Ring. Part of the difference is that the ring particles here are dominated by methane ice, in contrast to the water ice found in orbit around Saturn.

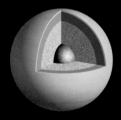

Uranus's outer layers of hydrogen, helium and methane form an atmosphere that changes from gas to liquid deeper inside the planet. This surrounds a mantle of churning ices, and a core of rock and ice.

Distance from sun	Orbital period	Orbital eccentricity	Diameter	Surface gravity	Rotation period	Axial tilt	Natural satellites
2.87	84	0.046	51,118	0.89	17.24	97.77	27
billion km	Earth years		km	g	hours	degrees	

2.87

billion km
from the Sun

Uranus

Active Uranus

Weather features

Although Uranus was disappointingly inactive during Voyager 2's 1986 flyby (see previous page), it seems that the weather on Uranus is highly seasonal, and when the Hubble Space Telescope took a look at the planet eleven years later, it revealed a world of obvious activity and bright storms. This false-colour picture also emphasizes the comparatively dark, narrow rings that encircle the planet.

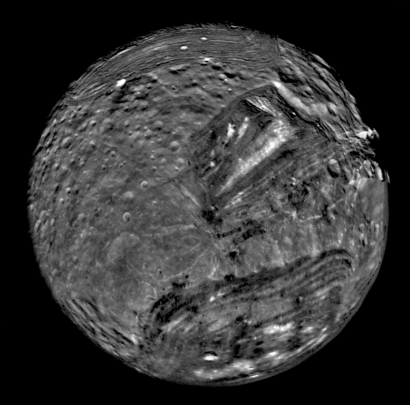

129.4

thousand km
from Uranus

Miranda

Diameter: 480 km

Icy natural satellite

Miranda is a tiny and strange patchwork moon. Surface features of varying ages and different origins are jammed up against one another with little rhyme or reason. At first astronomers thought that Miranda might have shattered to pieces and reformed after a cosmic collision, but it's more likely that the moon has frozen in the act of trying to develop a separate core, mantle and crust.

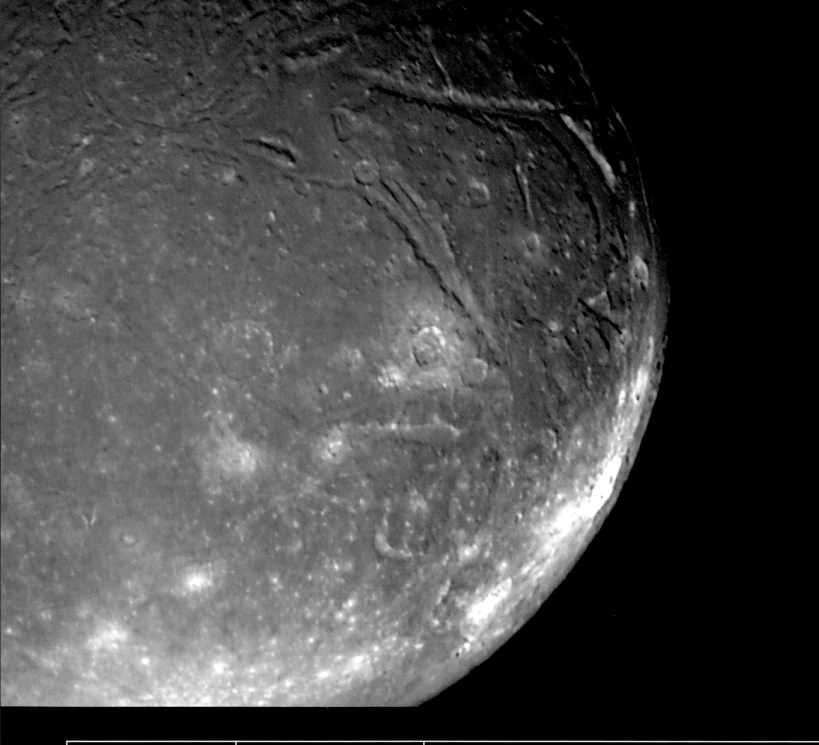

191

thousand km
from Uranus

Ariel

Diameter: 1,162 km

Icy natural satellite

The next moon out from Uranus, Ariel was large enough to develop properly. It has the youngest surface of any Uranian moon, suggesting that cryovolcanic activity on its surface in its early days resurfaced much of the moon with a fresh coating of ice. Ariel's standout feature, however, is Kachina Chasmata, a broad canyon system that runs across one hemisphere of the moon.

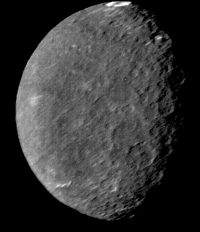

226.3
thousand km
from Uranus

Umbriel

Diameter: 1,169 km

Icy natural satellite

Ariel's near-twin in terms of size is somewhat further away from Uranus. Its surface is darker than its inner sibling, and it has a heavy coating of craters, suggesting that it has seen little geological activity since its formation. Umbriel is probably too far from Uranus to have benefited from tidal heating, and too small to have generated much internal heat as it formed.

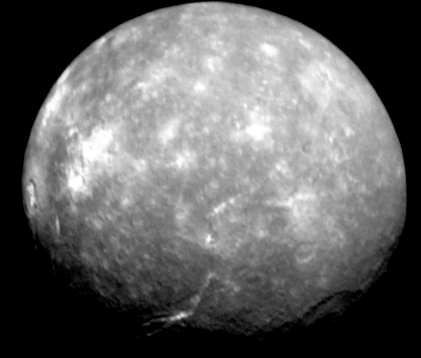

435.9

thousand km
from Uranus

Titania

Diameter: 1,578 km

Icy natural satellite

The largest of the Uranian satellites is Titania. This moon's surface, like Ariel's, shows some signs of resurfacing activity in its early days. While Ariel probably got the heat to drive such geological activity from tidal interaction with Uranus, Titania was large enough and rocky enough to develop an internal heat source without external assistance.

583.5

thousand km
from Uranus

Oberon

Diameter: 1,523 km

Icy natural satellite

The outermost major moon of Uranus, Oberon is slightly smaller than Titania, and has had a somewhat less active history. Voyager 2's distant photos of the moon revealed unusual craters with bright ejecta and a dark centre – probably caused when major impacts punctured the moon's crust and dark, carbon rich ice welled up from beneath.

Neptune – the Big Blue

Today, most astronomers would agree that Neptune is the outermost major planet. A near-twin of Uranus, discovered in 1846 from a mathematical study of wobbles in its inner sibling's orbit, it is a slightly smaller, somewhat bluer 'ice giant' world.

If Uranus was an obscure mystery world before the Voyager 2 flyby of 1986, then Neptune guarded its secrets even more closely. No telescope of the time was powerful enough to see any features on the disc, and only two satellites were known – the comparably large Triton, a world similar in size to some of the mid-sized satellites around Saturn, and the much smaller Nereid, with a wildly eccentric orbit that takes almost one Earth year to circle Neptune. Why, astronomers wondered, did this last giant have such a paltry system of moons compared to its inner neighbours?

Neptune's rings were another mystery – by the time Voyager 1 discovered the thin plane of dust around Jupiter in 1979, it seemed that ring systems of some kind were the rule rather than the exception around giant planets. But attempts to spot the rings of Neptune using the same technique of stellar eclipses that revealed the Uranian system proved frustratingly inconclusive. Sometimes a star would 'blink' on and off as though obscured by narrow rings, but on other occasions another star would pass close to Neptune with no apparent effect on its light. So by the time Voyager 2 approached its last planetary target in 1989, it had a number of questions to answer.

One thing no one had expected, however, was much sign of activity. With a cloudtop temperature even colder than that on Uranus, it was thought that the energy from the Sun would be far too weak to power any weather systems. But to everyone's surprise, the Neptune that eventually loomed out of the darkness showed bands reminiscent of Saturn, and a huge dark storm resembling Jupiter's Great Red Spot. Bright white high-altitude clouds known as 'scooters' also streamed around the planet, and when their movement was measured, it was discovered that Neptune had the strongest winds in the Solar System, capable of reaching speeds of up to 2,000 kph (1,250 mph).

The energy required to drive such vigorous weather systems clearly had to come from within the planet – but what could be generating it? All the giant planets, with the exception of Uranus, seem to have a similar internal power source, and the best theory to explain them is that energy is generated through friction as heavier materials continue to sift towards the centre of the planet, billions of years after the main era of planet formation came to an end. On Neptune, however, there is another twist. Like Uranus, it is an 'ice giant' – below a relatively thin gaseous atmosphere, it has a 'mantle' of slushy chemical 'ices', not only frozen water, but also frozen ammonia, methane, and other compounds. At a certain depth within the mantle, the temperatures and pressures must become high enough to break the methane apart into its chemical elements, carbon and hydrogen. In theory, such pressure would also be great enough to compress and bond the carbon atoms together, forming microscopic particles of the hardest substance known to man – diamond.

As for Neptune's other mysteries, Voyager 2 did a good job of solving them – fortunately, since no other probes to this remote world are planned for the near future. The images in the following pages reveal many of the solutions and capture the drama and spectacle unexpectedly lurking so far from the Sun.

Top: Voyager 2 discovered half a dozen new moons orbiting close in to Neptune, of which the largest and first to be found was Proteus, 440 km (270 km) long.

This blurry image is the best of Nereid yet obtained. Its distance from the planet varies between 817,200 km (507,500 miles) and 9.5 million km (5.9 million miles), yet it is believed to be one of Neptune's original satellite family, flung into its current orbit by the cataclysm of Triton's arrival (see p.208).

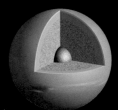

Neptune has a similar interior to Uranus, with an outer atmosphere of hydrogen, helium and methane, a deep mantle of chemical ices, and a small core made from a mixture of rock and ice.

Distance from sun	Orbital period	Orbital eccentricity	Diameter	Surface gravity	Rotation period	Axial tilt	Natural satellites
4.5	164.9	0.010	49,532	1.13	16.11	28.32	13
billion km	Earth years		km	g	hours	degrees	

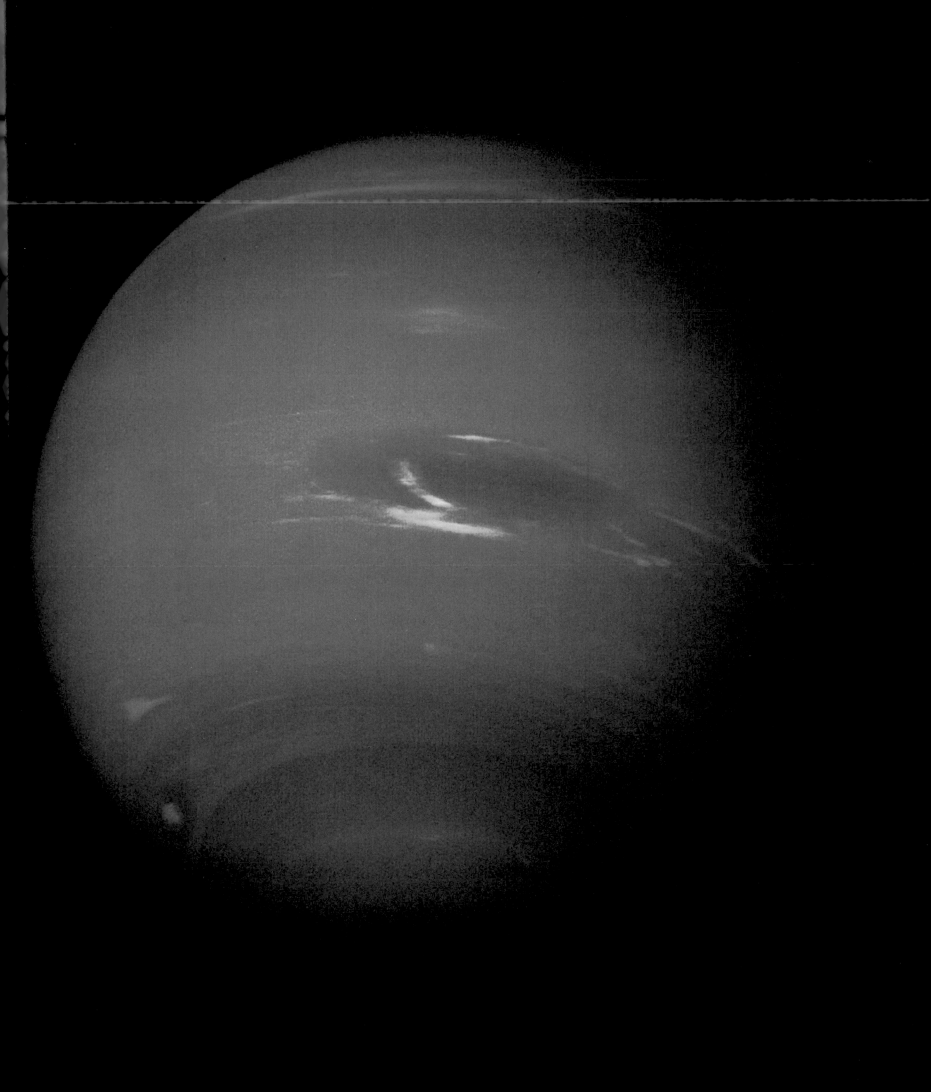

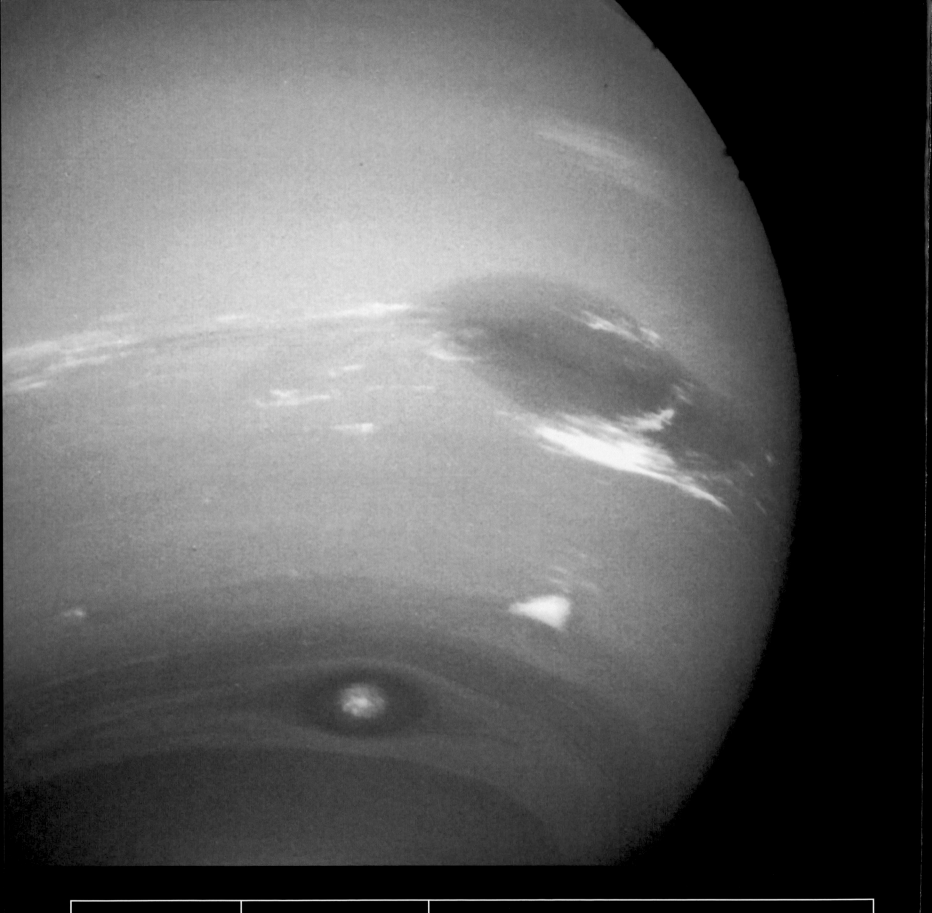

4.5

billion km
from the Sun

Neptune

Great Dark Spot and scooters

Weather features

Compared to Uranus, Voyager 2 found a wealth of activity when it arrived at Neptune. The most prominent feature was the so-called Great Dark Spot, a dark storm roughly the size of the Earth. Higher in the atmosphere are the white, roughly arrow-shaped cloud formations called scooters, which circle the planet at higher speeds than the lower cloud layers.

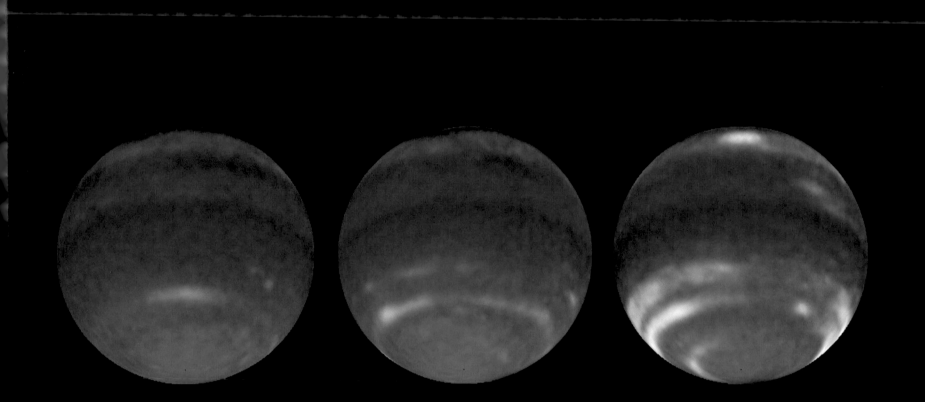

4.5

billion km
from the Sun

Neptune

Storms and belts

Weather features

Time-lapse images from the Hubble Space Telescope and the Hawaii-based Infrared Telescope Facility capture Neptune's features through three rotations, a decade on from Voyager's flyby. Astronomers were surprised to find that the original Great Dark Spot soon disappeared, but a new one has appeared, while the regions with the brighter high-altitude clouds remain the same.

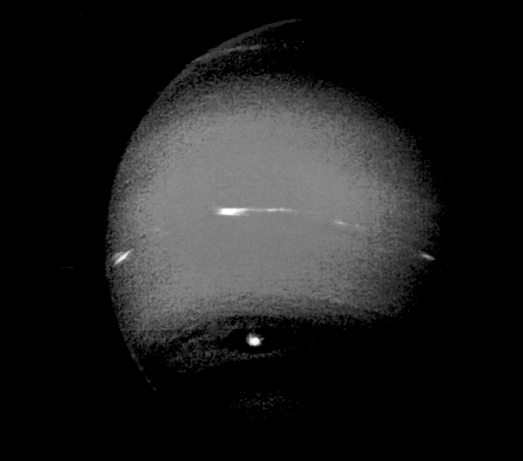

4.5

billion km
from the Sun

Neptune

Haze

Atmospheric feature

Both Neptune and Uranus get their colour from small amounts of methane in their atmospheres. Methane absorbs red light, hence the greenish colour of Uranus and the blue colour of Neptune, where the gas is more plentiful. Using filters designed to highlight the transparency in Neptune's atmosphere, Voyager 2 revealed the existence of a high-altitude haze around the planet.

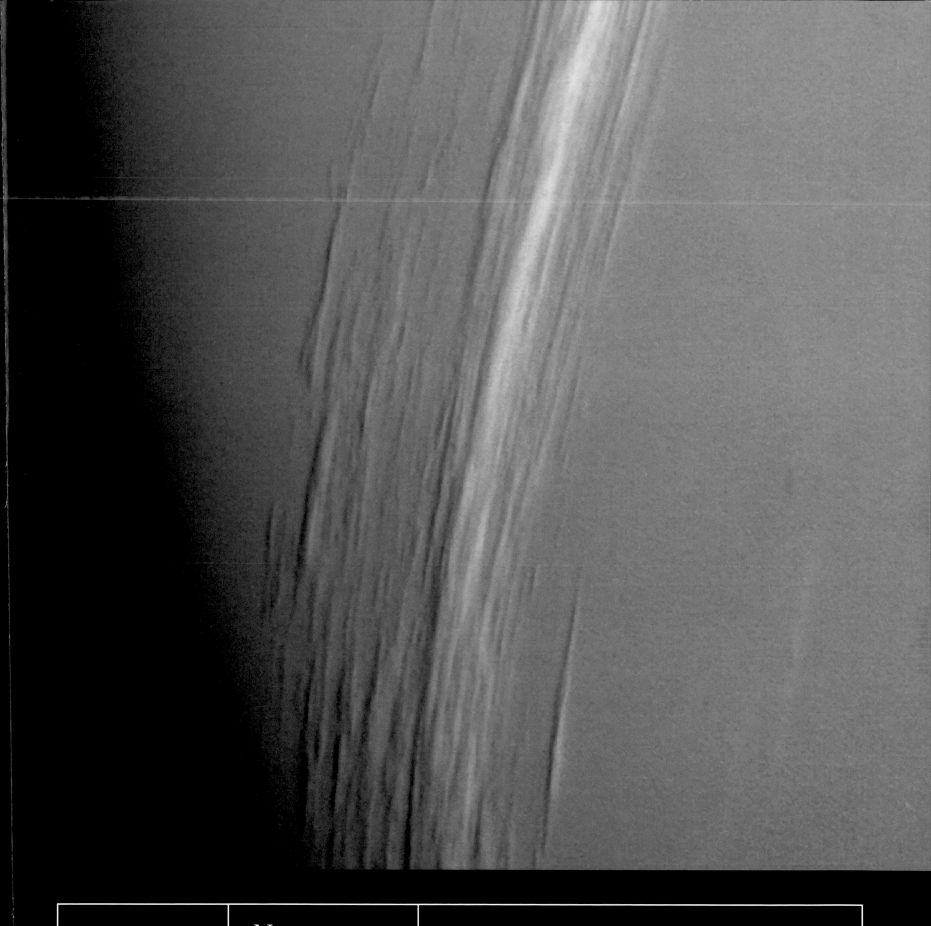

Neptune

High-altitude clouds

Weather feature

Neptune's high winds stretch out many of its clouds into long streamers. This deceptively placid-looking image from the Voyager 2 spaceprobe shows high-altitude bright cloud lanes casting their shadows onto the uniform blue of the atmosphere below. In reality, the winds around these clouds are moving at near-supersonic speeds.

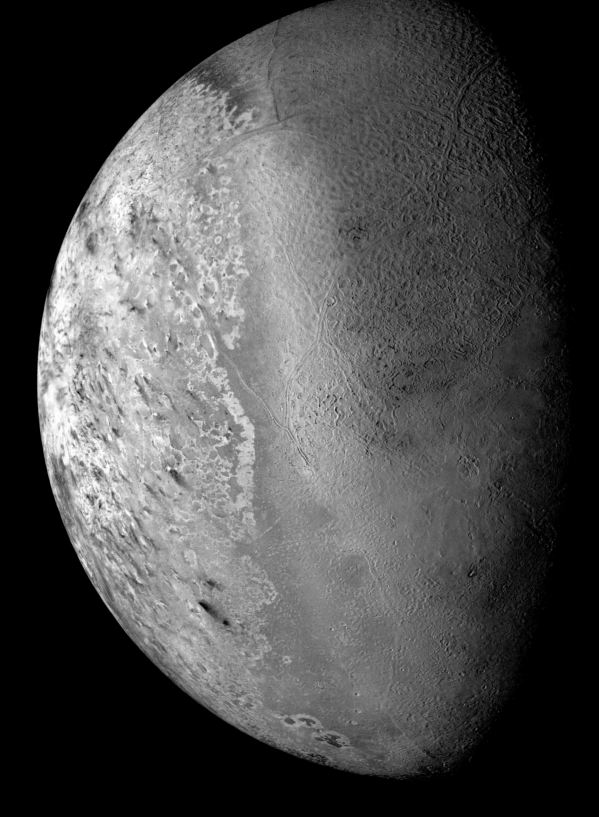

354.8

thousand km
from Neptune

Triton

Diameter: 2,707 km

Captured ice dwarf satellite

Neptune's major moon Triton has a surface divided into different types of terrain – bluish and hummocky 'cantaloupe terrain' in the north, and a smoother landscape, coated with grey and brown ices, in the south. Their origins lie in Triton's strange history – it seems to have originated elsewhere in the outer Solar System, before being captured by Neptune.

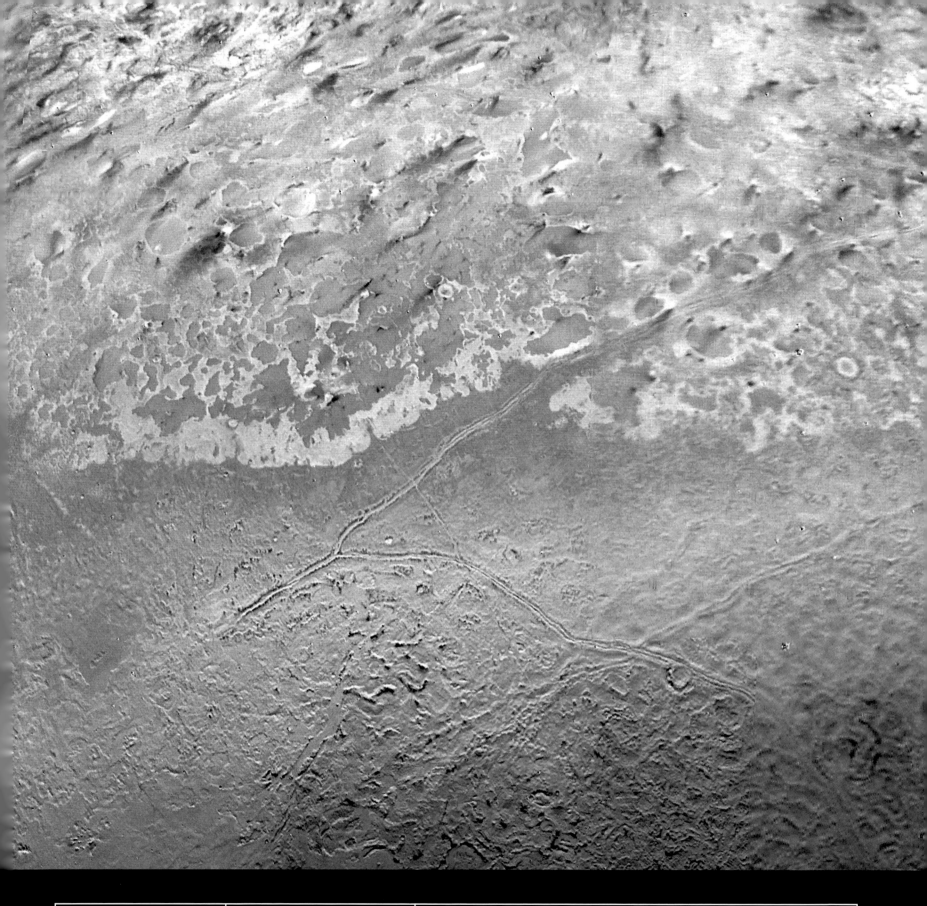

354.8

thousand km
from Neptune

Triton

Ice geysers

Cryovolcanic feature

Dark streaks across the smooth southern terrain are in fact shadows cast by geysers that belch a mix of nitrogen gas and dust into Triton's thin atmosphere. Geysers were a shock discovery on a world with a surface temperature around -235 °C (-391 °F). They are probably driven by tidal heat generated as Triton's orbit gradually spirals inward, dragging the moon to its eventual doom.

Ice Worlds

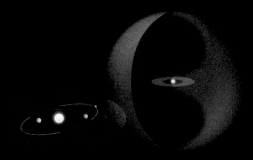

Even as you read these pages, the fastest spaceprobe ever built is shooting across the Solar System on a race against time. Its destination lies at the outer edge of the Sun's empire: the primary target is a flyby of tiny Pluto, for a long time considered the ninth planet of the Solar System. After that, the plan is to rendezvous with one or more of Pluto's cousins – the 'ice dwarf' worlds of the Kuiper Belt.

Launched in early 2006, the New Horizons probe will use a series of gravitational slingshot manoeuvres to steal momentum from the planets as it passes, ultimately receiving a large boost from Jupiter that will send it on a direct route to Pluto. If all goes well, it will reach this icy outpost in 2015. The probe's weight has been deliberately kept down to a comparatively light 480 kg (1054 lb). By using a powerful rocket at launch, the probe was already travelling at 16.21 km/s (36,300 mph) by the time it passed the orbit of the Moon.

Speed is of the essence because New Horizons is on a race against time. Studies of Pluto's light have revealed that it currently has a thin atmosphere, most likely made up of gases that have evaporated directly from the ice on the surface. It's thought that this atmosphere only forms for a few decades in Pluto's 248-year orbit, when it comes closest to the Sun and is briefly closer in than Neptune. But the last close approach happened in 1989, and this dwarf world is currently retreating from the Sun, its atmosphere slowly freezing back to the surface. In order to see Pluto with its ephemeral atmosphere in place, New Horizons needs to get there as soon as possible.

What will it find when it arrives? Pluto has always been a mystery, not just because of its distance but also because of its size – considerably smaller than Mercury, it was only considered a planet for so long by historical accident, thanks to the lucky discovery made by Clyde Tombaugh in 1930. At the time, Tombaugh was indeed searching for a new planet beyond Mercury, widely suspected by astronomers who thought there were unexplained 'wobbles' in Neptune's orbit. But within months of Pluto's discovery, astronomers had worked out that it was too small to affect Neptune, and eventually new calculations explained Neptune's orbit without any need for further large planets. Despite its

awkward characteristics, Pluto was classed as a planet partly because there was nothing else to compare it to. It was many decades before further 'Kuiper Belt Objects' were found, forming a roughly doughnut-shaped belt beyond Neptune.

Even though astronomers have studied Pluto for more than 70 years, they know surprisingly little about it. But it does have three moons, the largest of which, Charon, is almost half the size of the planet itself. This satellite is so large and so close to Pluto itself that tidal forces in this system have cut both ways – while Charon's rotation has slowed so it keeps one face permanently towards its parent, Pluto also rotates in the same period Charon takes to orbit. Technically, this makes Pluto and Charon a 'binary world' rather than a primary body–satellite system. The satellite orbits have also revealed that Pluto is tilted over on its axis even more than Uranus, by an angle of 122 degrees.

Physically, Pluto and Charon probably bear a passing resemblance to Neptune's satellite Triton, itself a suspected refugee from the Kuiper Belt. However, the tidal forces produced when Neptune captured its rogue satellite have reshaped the moon and given it a lot more activity than we might expect to see at Pluto.

As New Horizons sails on into the depths of the Kuiper Belt, it will encounter other Pluto-like worlds on its one-way trip out of the Solar System. However, it will miss Eris, the object larger than Pluto whose discovery forced astronomers to reconsider their definition of a planet, and ultimately led to the demotion of Pluto in 2006. It may also come across the dormant nuclei of 'short-period' comets dawdling out here at the slow-moving, lazy end of their orbits around the Sun. If we are lucky, we may still be detecting its radio signal in another decade or more, as it crosses the heliopause where the solar wind falters under the pressure of particles streaming out from other stars. The probe's power will dwindle to nothing as it crosses the void beyond the Kuiper Belt, the realm of Sedna and perhaps a few other objects in long elliptical orbits. And it will surely be a cold, inactive relic by the time it reaches the motherlode of comets – the vast spherical shell of the Oort Cloud, around one light year from Earth.

The Kuiper Belt (above left) stretches from the edge of Neptune's orbit to at least 12 billion kilometres (7.5 billion miles) from the Sun. On the right, the Oort Cloud surrounds the entire Solar System at a distance of one light year (9.5 trillion km, or 5.8 trillion miles). In this illustration, the Kuiper Belt would be little more than a pinhead if shown to scale.

A typical ice dwarf has a thin crust made from various ices, wrapped around a mostly water-ice mantle. They are also thought to have a large core made mostly of rock.

billion km
from the Sun

Pluto

Diameter: 2,304 km

Ice dwarf

Even the clearest direct images of Pluto from Earth show it as a starlike point of light. However, a Hubble Space Telescope photo from 2006, the clearest yet, reveals a surprise – as well as Pluto itself and its large companion Charon, it showed two more small satellites, named Nix and Hydra. These new moons are around 50 km (30 miles) across, and orbit Pluto further out than Charon.

5.7–14.6

billion km
from the Sun

Eris

Diameter: 2,400 km

Ice dwarf

Discovered to be just larger than Pluto, Eris – or 2003 UB₃₁₃ as it was originally labelled – ignited the whole argument over how to define a 'planet' that ultimately led to the demotion of Pluto from the planetary club. Appropriately named after the Greek goddess of discord, Eris has an unusually bright surface, and an orbit that takes it much further away from the Sun than its rival.

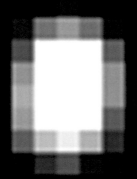

billion km
from the Sun

Quaoar

Diameter: 1,300 km

Ice dwarf

The first substantial world to be discovered in the Kuiper Belt after Pluto, Quaoar is considerably larger than Ceres, the largest asteroid, and about the size of Pluto's moon Charon. Such tiny worlds are impossible to distinguish directly from stars through a telescope – they can only be discovered by their slow movement across the more distant background of stars.

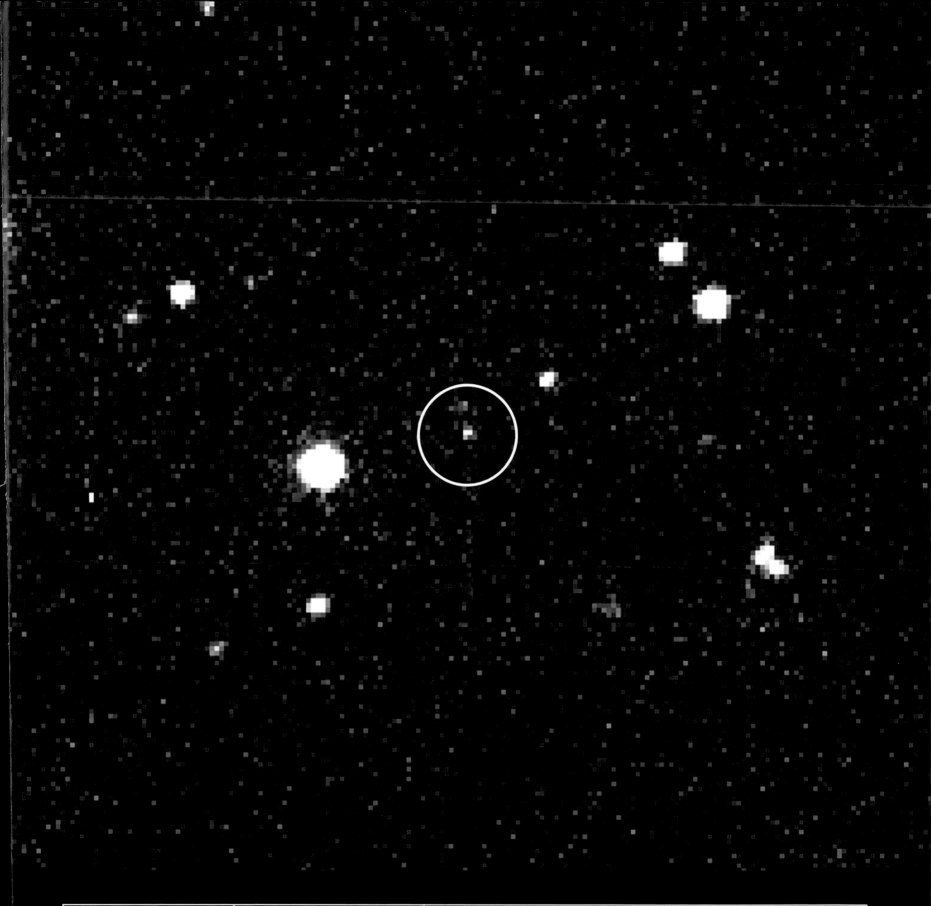

11-148
billion km
from the Sun

Sedna

Diameter: 1,500 km

Ice dwarf?

The most distant known object in the Solar System is Sedna, a mysterious world that lurks between the Kuiper Belt and the Oort Cloud in a 10,500-year orbit. Named for an Inuit goddess of the Arctic Ocean, little is known about this new world, except that it has a remarkably red surface – almost as red as the landscape of Mars.

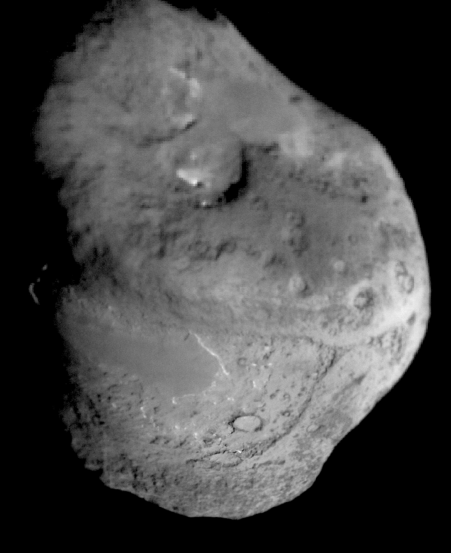

225
million km
from Sun at perihelion

Tempel 1

Length: 7.6 km

Comet

Comet Tempel I's 5.5-year orbit made it an ideal target for NASA's Deep Impact probe, which fired a barrel-like projectile into the surface and studied the fountain of material that the impact blasted into space. Surprisingly, the comet's material seems to have gone through a number of chemical changes – it was not the pristine iceball from the early Solar System that was expected.

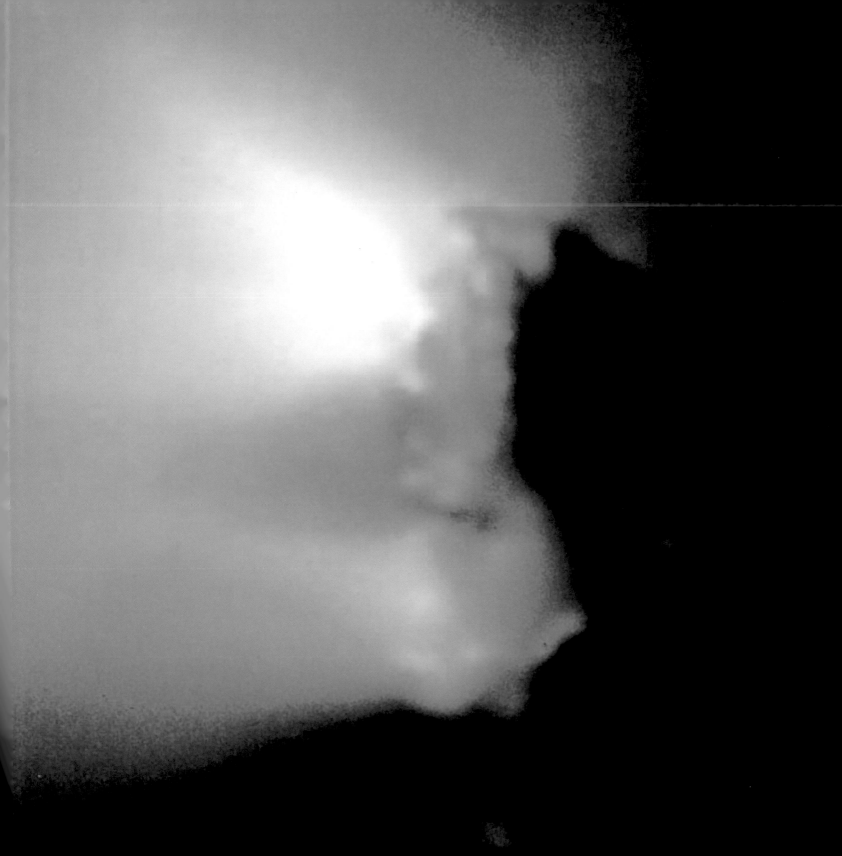

88

million km
from Sun at perihelion

Halley

Length: 16 km

Comet

The first good view of a comet nucleus came from the Giotto probe to Halley's Comet in 1986. Halley is a relatively recent addition to the inner Solar System that returns every 76 years and has been recorded by humans since at least 240 BC. Halley's youth means that it has plenty of fresh ice just beneath its surface, and so it is the most active of the short-period comets.

144
million km
from Sun at perihelion

NEAT

Diameter: unknown

Comet

One of the best naked-eye comets of recent years, Comet NEAT was discovered by NASA's Near-Earth Asteroid Tracking project in 2001, three years before it passed closest to the Sun. As it swung inside Earth's orbit, it developed a large coma more than twice the size of Jupiter, and a tail millions of kilometres long. Only the innermost regions of this tail are shown in this photo.

114
million km
from Sun at perihelion

Linear

Diameter: unknown

Comet

As a comet passes perihelion and comes closest to the Sun, it can experience a great deal of stress, due to heat and tidal forces. Sometimes this can cause the comet to break up completely, as happened when Comet Linear rounded the Sun in 2001. The Hubble Space Telescope found a cloud of mini-comets, debris from the disintegration, following the orbit of their departed parent.

Glossary

Aphelion
The point in a Solar System object's orbit at which it is furthest from the Sun.

Arachnoid
A depressed network of radial and concentric cracks in the surface of Venus, thought to be formed as the surface bulges out under pressure from magma trapped beneath, and then subsides as the magma retreats or escapes elsewhere.

Asteroid
A small rocky world orbiting in the inner Solar System. Nearly all asteroids are irregular in shape, and most orbit within the 'asteroid belt', confined between the orbits of Mars and Jupiter.

Atmosphere
A thin shell of gases trapped around a planet by its gravity. Atmospheres vary from the substantial and opaque (around worlds such as Venus and Titan) to the sparse and invisible (the atmosphere around Europa). The outer layer of a giant planet, below which gas condenses to liquid or ice, is also called the atmosphere.

Aurora
An atmospheric phenomenon seen on planets with substantial magnetic fields. Particles from the solar wind are trapped in the planet's magnetic field, and funnelled towards the magnetic poles. As they collide with gas atoms and molecules, they emit visible light and radio waves.

Axial tilt
The amount by which a planet's north pole is tilted over from 'straight up' relative to the plane of its orbit. Because a planet's north pole is defined by its rotation (the planet rotates counterclockwise as seen from above the north pole), it is possible for planets to have axial tilts of more than 90 degrees if the north pole is effectively pointing 'downwards'.

Caldera
A circular depression or crater around a volcanic vent.

Coma
The vast but tenuous 'atmosphere' of escaping gases that grows around a comet's nucleus as it approaches the Sun.

Comet
A small icy world from the outer Solar System, originally formed in the region around the orbits of Uranus and Neptune, but later flung out into the distant Oort Cloud. Occasionally, comets come plunging back into the inner Solar System, where they may develop a coma and tail as they heat up. A few become trapped in orbits much closer to the

Core
The central region of a planet whose internal layers have separated through differentiation. Planetary cores are usually rich in metals and other heavy elements – the cores of most terrestrial planets are thought to contain mostly nickel and iron, and the largest are still molten, partly or completely.

Corona
A depressed pit on the surface of Venus surrounded by a pattern of concentric rings. Coronae are thought to have a similar origin to arachnoids.

Crater
A roughly circular depression in a planet's crust, often surrounded by a raised rim. Craters can be caused by meteor impacts or volcanic action.

Crust
The exposed outer surface of a terrestrial planet or moon. In properly differentiated planets, the crust is a distinct layer, perhaps a few tens of kilometres deep, above the mantle and core. In other cases, the term crust refers to the planet's surface itself.

Cryovolcanism
A form of volcanism fuelled not by molten rock, but by a viscous mixture of water and ammonia ices that can stay fluid in the low temperatures of the outer solar system, and which shares many of the properties of volcanic lava.

Differentiation
The process in which a planet or moon's interior separates into distinct layers. On solid worlds, it only occurs if the interior grows hot enough to melt and then separate as gravity pulls denser material to the centre. In general, this only occurs in the Solar System's larger worlds.

Eccentricity
A measure of how 'stretched' a planet or moon's elliptical orbit is. A perfect circle has an eccentricity of 0, and as the ellipse becomes more stretched, its eccentricity approaches, but never reaches, 1.

Ejecta
The pulverised material flung out from an impact site during the formation of a crater. Ejecta contains a mix of subsurface rock and fragments of the colliding body. It typically forms a blanket close to the crater, but can be sprayed over much longer distances as distinct rays.

Fault
A crack in the otherwise continuous crust of a moon or planet. On Earth, most major faults occur along the boundaries of tectonic plates, but faults can also occur where the crust is stretched or pulled in opposite directions by movements in the mantle.

Giant planet
A planet many times larger than the terrestrial worlds, but composed of far less dense material. The giant planets are Jupiter, Saturn, Uranus and Neptune, but they are sometimes split into 'gas giants' (the inner pair, largely composed of light gases that turn to liquids in their high-pressure interiors), and 'ice giants' (the outer pair, with a relatively thin atmosphere surrounding a mantle of slushy chemical ices).

Granulation
A dark cellular pattern on the surface of a star like the Sun, formed at the tops of convection cells, where hot material rises in the centre and cool material sinks around the edges.

Ice
In chemical terms, an ice is the solid form of any 'volatile' chemical (i.e., one with a relatively low boiling point). Common ices in the Solar System include water, nitrogen, methane, carbon monoxide and dioxide, and ammonia.

Ice dwarf
A small 'dwarf planet', usually found in the Kuiper Belt beyond Neptune, and composed mostly of ice with some rock. Pluto is the most famous ice dwarf, and Neptune's satellite Triton is another.

Impact Basin
A large impact crater that has been changed since its formation, often due to flooding with material erupted from inside the planet.

Impact Crater
A crater formed by a meteorite impact. Unlike volcanic calderas, there is no lower limit to the size of an impact crater. Small ones create simple, bowl-shaped depressions, but larger examples can have terraced walls, and mountainous central peaks.

Infrared
Electromagnetic radiation emitted by warm objects, usually too cool to generate visible light.

Kuiper Belt
A doughnut-shaped ring of ice dwarf worlds surrounding the planetary region of the Solar System. Many 'short-period' comets (those with orbits lasting less than 200 years) reach their aphelion in the Kuiper Belt.

Lava
Molten rock that has erupted onto the surface of a planet or moon. Cryolava is a low-temperature equivalent, a slushy mix of ammonia and water ice.

Magma
Molten rock that has risen from a planet's mantle or been formed by heating in a planet's crust, but has not yet erupted

Magnetic field
A force field that surrounds any object and influences other susceptible objects ranging from subatomic particles to metal spacecraft. Magnetic fields in planets are generated by the large scale movement of electrically conducting material, such as the molten metal in the core of a rocky planet, or the liquid hydrogen mantle inside a gas giant.

Mantle
The intermediate layer inside a differentiated planet, between the core and the outer crust (in a terrestrial planet) or atmosphere (in a giant planet). In larger worlds, the mantle may be semi-molten and constantly shifting, which may drive changes at the surface such as tectonics.

Meteor
A 'shooting star' – the glowing trail made by a particle from space entering a planet's atmosphere and heating up due to friction.

Meteorite
A large particle from space that makes it through a planet's atmosphere and hits the ground.

Meteoroid
A catch-all term for any potential meteor or meteorite – a small object orbiting in space, ranging from dust particles to large boulders, and blending at the upper end of the scale with small asteroids.

Moon
A moon (with a small 'm') is any satellite with a natural origin in orbit around a planet.

Nucleus
For most of a typical comet's orbit, it consists of nothing but the nucleus, an icy solid body with a thin crust of dark, carbon-rich material. The dark surface increases the amount of heat the nucleus absorbs as it comes closer to the Sun, allowing the ice below to heat up, and vapour to escape to form the comet's coma and tail.

Near Earth Asteroid (NEA)
An asteroid whose orbit brings it closer to Earth than the main belt. Not all NEAs have potentially dangerous, Earth-crossing orbits.

Oort Cloud
A vast spherical shell of comets orbiting the Solar System at a distance of about 1 light year, around the limit of the Sun's gravitational reach. The cloud contains comets that formed much closer to the Sun, but were expelled by encounters with the giant planets in the early days of the Solar System. Occasional disturbances can send comets falling back on a long elliptical path towards the Sun.

Palimpsest
A large, bright-floored crater found on Ganymede or Callisto. Palimpsests get their name from the word for a reused scroll or parchment, and are regions where major impacts penetrated the moon's crust, allowing fresh ice from the mantle to flood the crater floor.

Perihelion
The point in a Solar System object's orbit at which it is closest to the Sun.

Radiation
In astronomy, radiation usually refers to electromagnetic (e-m) radation, a form of electrical and magnetic disturbance that is emitted by most objects in the Universe. Visible light is a type of e-m radiation, but objects must be quite hot to emit it (most everyday objects, including the planets, only shine by reflecting visible light). Lower-energy forms of radiation include infrared and radio waves, while higher-energy forms include ultraviolet, X-rays and gamma rays. 'Ionising radiation' is the correct term for the particles and high-energy gamma rays associated with radioactivity.

Ring system
A dense stream of particles in orbit around a (usually giant) planet, which remain confined to a flat plane and perfectly circular orbits in order to avoid collisions with each other. Ring systems range from the vast halos around Saturn to the thin, clumpy structures orbiting Neptune.

Satellite
Any object in orbit around another, but usually used as a synonym for moon. The term 'natural satellite' distinguishes moons formed at the same time as the planet they orbit from 'captured satellites', which are typically asteroids or comets caught by the planet's gravity later in its history.

Shield volcano
A structure formed over the course of many eruptions as lava builds up around the original volcanic vent, typically forming a volcano with shallow slopes, spread across a huge area.

Solar System
The spherical region of space, roughly 3 light years or 28.5 million million km (17.7 trillion miles) across, in which the Sun's gravity is dominant.

Solar Wind
A stream of particles flowing from the Sun. The solar wind extends throughout the inner solar system to well beyond the orbit of the Kuiper Belt, where it finally slows and falters under the inward pressure of the winds from other stars.

Star
A ball of gas dense enough to generate energy in its core by fusing the atomic nuclei of lighter elements to make heavier ones. The Sun and most stars in the sky generate energy through fusion of hydrogen into helium.

Sulcus
A broad swathe of parallel ridges across the surface of Ganymede, thought to have formed as the giant moon's icy surface became stretched and cracked allowing fresh ices to well up from within.

Synchronous orbit
An orbit in which a satellite spins on its axis in the same time that it takes to orbit its planet, so that it keep one face permanently turned towards the planet. Most moons in the Solar System (including our own) have synchronous orbits.

Tail
A stream of particles emanating from a comet's coma and stretching across interplanetary space. Comets frequently have both a gas and a dust tail. The bluish gas tail is always blown away from the Sun, regardless of which direction the comet is moving, so it often travels in front of the comet itself. The dust tail is also blown by the solar wind, but tends to stay closer to the track of the comet's orbit.

Tectonic plate
A large chunk of a planet's crust floating independently on top of the mantle, which may separate from, grind past, or collide with, other plates. The only planet with fully developed tectonic plates is Earth.

Tectonics
A process that shapes the appearance of terrestrial planets and moons through the slow, inexorable drift of a solid crust on top of a semi-molten, mobile mantle. Earth has the most developed tectonics of any planet, but evidence of past tectonics can be found on worlds ranging from Venus to Io.

Terrestrial planet
A rocky planet, essentially resembling Earth. The four rocky planets are Mercury, Venus, Earth and Mars.

Tidal heating
A gravitational effect that heats the interiors of satellites orbiting close to giant planets. As the moon's distance from the planet changes slightly throughout each orbit, the gravitational pull on it changes, and the planet flexes and changes its shape. Friction in the interior can be enough to melt the moon's core and trigger geological activity such as volcanism or cryovolcanism.

Ultraviolet
Electromagnetic radiation with shorter wavelengths than visible light, emitted only by particles with high energies such as the surfaces of very hot objects. Within the Solar System the Sun is the dominant source of ultraviolet light.

Index

Picture credits

Quercus
21 Bloomsbury Square
London
WC1A 2QA

Copyright © Quercus Publishing Ltd 2006

Book design: Grade Design Consultants, London
www.gradedesign.com

ISBN: 1-905204-26-4